Biodiversity and Biogeographic Patterns in Asia-Pacific Region I: Statistical Methods and Case Studies

Authored By

Youhua Chen

Department of Renewable Resources
University of Alberta
Edmonton
T6G 2H1
Canada

CONTENTS

Biodiversity and Biogeographic Patterns in Asia-Pacific Region I: Statistical Methods and Case Studies

FOREWORD

Biogeography is the biological discipline that studies the geographic distribution of plant and animal taxa and their attributes in space and time. Nowadays, it is passing through a revolution concerning its foundations, basic concepts, and methods. As part of this revolution, biogeographers are increasingly recognizing the need of integration with other disciplines, in order to develop a truly interdisciplinary and pluralist science. The recent discipline of conservation biogeography shows clearly the possibilities of interdisciplinary collaboration. Additionally, during the last decades, there has been a considerable progress in the quantitative analysis of biogeographical patterns, with statistical multivariate and phylogenetic methods proposed to analyse particular ecological and biogeographical patterns and to establish meaningful comparisons.

The book written by Youhua Chen presents a thoughtful analysis of different biogeographical methods and their application to the analysis of biotic patterns in the Asia-Pacific region, which encompasses China, India and southeast Asia, and represents a very interesting region, which possesses an outstanding biodiversity. The analyses of different plant and animal taxa, using several biogeographical methods, led Youhua to identify clear patterns and to correlate them with climatic and biological data. Finally, he highlighted the relevance of these patterns for biodiversity conservation.

There are different reasons to read this book. Biogeographers will be interested in an outstanding region of the world, with specific chapters on birds and plants as case studies. Conservationists will appreciate the identified biogeographical patterns, which might help them establish conservation priorities. Ecologists will have a modern review of several methods used to analyse species-area relationships, species abundance and modelling species' potential distributions, among other patterns. Naturalists will be delighted with the rich biological information conveyed in the analyses. Students will learn about theoretical issues, and the case studies will surely make them think about biogeographic questions and how to answer them. I am sure that all of them will enjoy this very interesting book!

Juan J. Morrone
Museo de Zoología 'Alfonso L. Herrera'
Departamento de Biología Evolutiva
Facultad de Ciencias
Universidad Nacional Autónoma de México (UNAM)
Apartado Postal 70-399, 04510 Mexico D.F., Mexico

PREFACE

In the past several decades, biodiversity conservation has gradually become a mainstream sub-discipline in contemporary biology. The importance of biodiversity conservation is straightforward for non-scientists to understand because in this post-industrialization and information-based era, people become more aware about the harmony between human being, society and nature. Biodiversity science is a leading multidisciplinary science in 21st century.

In tropical and temperate Asia, there are two BRICS countries: India and China, both are giant and developing countries. At the end of 2013, human population sizes of both countries rank at the top two across the nations of the whole world. Species extinction and biodiversity conservation in these areas is definitely challenging and urgent because of the expanding population, environmental pollution, rapid urbanization and industrialization.

In comparison to developed countries, biodiversity survey in most Asian countries seems not so comprehensive. New species are growingly discovered in the region, and the mega-biodiversity and original forests in the region provide invaluable resources for ecologists to explore biodiversity patterns over there. It is a great opportunity to present biodiversity and biogeography patterns at the region. My book can convey valuable information for researchers interested in the conservation issues of the Asian countries.

This book is dedicated to the quantitative analyses and systematic discussion of spatial biodiversity and biogeographic patterns in Asia-pacific region, including China, India and other south-eastern Asian countries. In the book, for the first part, the modern statistical and numerical methods have been introduced to the readers to tell them how to conduct spatial macro-biodiversity and biogeography analyses. These statistical methods have been widely used by researchers that are still active in the field of biodiversity and ecology. In second part of the book, different case studies over the region were provided, which covered very broad topics using many quantitative methods that have been introduced in the first part of the book, including phylogenetics, spatial statistics, multivariate statistics and others. As case studies, I provided detailed interpretation of the quantitative results and how these results are relevant to local and regional ecological processes. The book is suitable for anyone interested in biodiversity conservation to read. In particular, it is adequate for the undergraduate students that need a textbook to learn ecological methods and graduate students and scholars that need to know the recent advances in the field of biodiversity conservation, biogeography and macroecology.

The key features of my book are,

1. The book focuses on Asia-Pacific region, which is a very unique region and mega-biodiversity hotspot in the world. As far as I know, there is no book contributing to the biodiversity conservation knowledge over the region. Thus, my book is the first one that only focuses on biodiversity and biogeography in Asia-Pacific region systematically and comprehensively.

2. The book focuses on discussing the statistical methods on spatial biodiversity and biogeography patterns, thus may be of interests to academic researchers.

3. The statistical methods in my book cover a broad range and represent the most up-to-dated ones that are widely used in current biodiversity and macroecology studies, including spatial statistics, phylogenetic theory, neutral networks and ecological genomics.

The audience of the book includes university libraries, academic scholars and professors, graduate and undergraduate students.

Writing of the book has been generously supported by the China Scholarship Council. I am very grateful to Prof. Juan J. Morrone, one of the most eminent biogeographers, for kindly writing the foreword for my book and supporting my researches for a long time since we know each other. At last, this book is dedicated to my parents (Fashen Chen and Yuying Zhong) and my sisters (Ying, Fang and Yuan) for their love and support. Definitely, I can not complete this book without them behind my back.

ACKNOWLEDGEMENTS

This work is supported by the China Scholarship Council (No. 201308180004).

CONFLICT OF INTEREST

The author confirms that this eBook contents have no conflict of interest.

Youhua Chen
Department of Renewable Resources
University of Alberta
Edmonton
T6G 2H1
Canada
E-mail: haydi@126.com

Mathematical Operators

$\min(a,b)$: minimal value between a and b.

$\max(a,b)$: maximal value between a and b.

X^T : transpose of matrix X.

X^{-1}: inverse of matrix X.

$Tr(X)$: trace of a square matrix X.

$E(X)$: expectation of a random variate X.

$Var(X)$: variance of a random variate X.

$\tanh(X)$: hyperbolic tangent of a scalar value X.

$\overline{x}$: mean value of a vector x.

$\langle x \rangle$: mean value of a vector x, similar to $\overline{x}$.

$x!$: factorial for a non-negative integer, equal to $x = x \cdot (x-1) \cdot (x-2)..3 \cdot 2 \cdot 1$.

$\begin{pmatrix} m \\ n \end{pmatrix}$: binomial coefficient, equal to $\dfrac{m!}{(m-n)!n!}$.

$\Gamma(x)$: Gamma function, equal to $\int_0^\infty t^{x-1} e^{-t} dt$; when x is a non-negative integer, $\Gamma(x) = (x-1)!$

1

CHAPTER 1

Species Richness and Diversity

Abstract: This chapter provides some metrics for measuring species diversity, the most basic and important diversity component in ecological studies. The metrics for species diversity covered only some common used ones in this book, like Shannon and Simpson indices. The closed forms for computing the variance of the relevant indices are also provided. Typical methods for the extrapolation of species richness are also mentioned in the text.

Keywords: Ecological indicators, ecosystem functioning, extinction risk, functional diversity, functional redundancy, functional traits, molecular information, phenotype, statistical ecology.

INTRODUCTION

Species richness is simply the number of species in your ecological system. Species diversity is a bit complex, in addition to quantify the total number of species in the system, it also evaluated how even of the population sizes (or abundance) over different species in the system. Thus, diversity should be measured in two facets: count and evenness.

The concept of species diversity could be extended to taxonomic classes that are beyond the species concept. For example, at population level, diversity is usually called as genetic diversity. At taxonomic classification level, diversity is termed as taxonomic diversity and/or phylogenetic diversity, depending on the metrics that are used to quantify diversity. At last, at trait perspective, diversity is termed as functional diversity. Thus, biological diversity is a definition with tremendous meanings. Biological diversity constitutes the beautiful nature surrounding our humans on the earth.

In this Chapter, I will present some common statistical methods for quantifying species diversity. Also, the methods for extrapolating the diversity patterns will be presented. For other diversity components, like phylogenetic diversity and functional diversity, they will be presented in other sectors.

MEASURE OF SPECIES DIVERSITY

Species Richness

Species richness for a local sample is simply to count the number of species found in the sample. It's calculation reads,

Youhua Chen

$$S_{local} = \sum_{i=1}^{S} I(p_i) \tag{1}$$

Where $I(p_i)$ is an indicator function, $I(p_i)=1$ if $p_i>0$, and $I(p_i)=0$ if $p_i=0$.

Shannon Diversity Index

Shannon index [1] is the most classical index for measuring species diversity, its calculation reads,

$$H' = -\sum_{i=1}^{S} \frac{n_i}{N} \ln \frac{n_i}{N} \tag{2}$$

Where N is the total individuals found the sample from all the species. S is the total number of species in the sample. n_i is the abundance for species i.

The evenness of the Shannon index is computed as,

$$J' = \frac{H'}{H'_{\max}} \tag{3}$$

Where $H'_{\max}$ is computed as $H'_{\max} = -\sum_{i=1}^{S} \frac{1}{S} \ln\{\frac{1}{S}\} = \ln S$.

Thus, evenness is calculated as,

$$J' = \frac{-\sum_{i=1}^{S} \frac{n_i}{N} \ln \frac{n_i}{N}}{\ln S} \tag{4}$$

The variance of Shannon's index is given by,

$$\operatorname{var}(H') = \frac{\sum_{i=1}^{S} n_i \ln(n_i)^2 - (\sum_{i=1}^{S} n_i \ln(n_i))^2 / N}{N^2} \tag{5}$$

or,

$$\operatorname{var}(H') = \frac{\sum_{i=1}^{S} n_i \ln(n_i)^2 - (\sum_{i=1}^{S} n_i \ln(n_i))^2 / N}{N^2} + \frac{S-1}{2N^2} \tag{6}$$

The second one is more accurate [2], especially for small sample cases.

Simpson Diversity Index

Simpson's index *D* [3] might be the most meaningful measure of evenness. *D* is the probability that two randomly sampled individuals are from two different species. Such a definition is analogous to the genetic concept-heterozygosity.

$$D = 1 - \frac{\sum_{i=1}^{S} n_i(n_i - 1)}{N(N-1)} \tag{7}$$

where *N* is the total individuals found the sample from all the species. *S* is the total number of species in the sample. n_i is the abundance for species *i*.

For computing 95% confidence interval, the variance of the Simpson's index *D* should be known, which reads,

$$\mathrm{var}(D) = \frac{\sum_{i=1}^{S}(\frac{n_i}{N})^3 - \left\{\sum_{i=1}^{S}(\frac{n_i}{N})^2\right\}^2}{0.25N} \tag{8}$$

or a more accurate closed form as follows,

$$\mathrm{var}(D) = \frac{4N(N-1)(N-2)\sum_{i=1}^{S}(\frac{n_i}{N})^3 + 2N(N-1)\sum_{i=1}^{S}(\frac{n_i}{N})^2 - 2N(N-1)(2N-3)\left\{\sum_{i=1}^{S}(\frac{n_i}{N})^2\right\}^2}{(N(N-1))^2} \tag{9}$$

The above formula is said to very suitable for small sample cases [2].

Renyi Entropy

The Rényi entropy is a generalization of the Shannon entropy to other values of *q* than unity. It can be expressed:

$${}^{q}H = \frac{1}{1-q}\ln\left(\sum_{i=1}^{S}\left(\frac{n_i}{N}\right)^q\right) \tag{10}$$

q defines the order of the entropy.

EXTRAPOLATION OF SPECIES RICHNESS

Chao1 and Chao2 Indices

Chao1 and Chao2 indices [4] might be the most common used ones for extrapolating regional species richness based on the data from local samples (Chao2 index) or the species' richness in a fixed sample (Chao1 index, if we measure the abundance of each species in the fixed sample). Its calculation formula is given by,

$$S_{Chao} = S_{obs} + \frac{F_1^2}{2F_2} \tag{11}$$

If $F_2 = 0$, then $S_{Chao} = S_{obs} + F_1(F_1 - 1)/2$.

where S_{Chao} is the estimated richness at regional scale (or sample) or a fixed sample, S_{obs} is the number of species that are found across all the local samples in the fixed sample, F_1 is the number of singletons (*i.e.*, the number of species with only one individual in the fixed sample or the number of species that are found in only one local sample across all the samples at regional perspective) and F_2 is the number of doubletons (the number of species with only two individuals in the fixed sample or the number of species that are found in only one local sample across all the samples if regional richness is extrapolated).

The simple idea behind the estimator is that if a sampled site contains a lot of rare species that are not found in other sites (that is, they are only distributed in the focused site, being singletons), it is very likely that there are more rare species in the site required to be detected or discovered.

To get the 95% confidence interval for the Chao1 estimator, the variance of the index should be computed, which followed [5],

$$\operatorname{var}(S_{Chao}) = \frac{F_1(F_1 - 1)}{2(F_2 + 1)} + \frac{F_1(2F_1 - 1)^2}{4(F_2 + 1)^2} + \frac{F_1^2 F_2 (F_1 - 1)^2}{4(F_2 + 1)^2} \tag{12}$$

To get lower and upper bounds of the 95% confidence interval, the following statistics should be calculated,

$$C = \exp\left\{1.96\sqrt{\ln(1 + \frac{\operatorname{var}(S_{Chao})}{(S_{Chao} - S_{obs})^2})}\right\} \tag{13}$$

$$LCI_{95\%} = S_{obs} + \frac{S_{Chao} - S_{obs}}{C}$$
$$UCI_{95\%} = S_{obs} + C(S_{Chao} - S_{obs}) \tag{14}$$

where $LCI_{95\%}$ and $UCI_{95\%}$ represent the lower and upper bounds of the 95% confidence interval.

Jackknife Estimators

The Jackknife estimator [6] is also commonly used for estimating regional richness (not abundance-based), which is given by,

$$S_{jack1} = S_{obs} + \frac{m-1}{m} F_1 \tag{15}$$

Where m denotes the total number of local samples. In particular, here, F_1 only denotes the number of species that are found in only one local sample across the whole region.

The above index is a first-order estimator; there is a second-order version [7] of the estimator, which is given by,

$$S_{jack2} = S_{obs} + \frac{2m-3}{m} F_1 - \frac{(m-2)^2}{m(m-1)} F_2 . \tag{16}$$

Rarefaction Curve

Rarefaction is also widely used to interpolate species richness. Let $S_{ind}(m)$ denotes the expected number of species from a random sample with m individuals which is drawn from a reference pool with S species and n individuals in a total ($n > m$) [8]. If the true probabilities/relative abundance of each species i p_i was known prior, then the species frequencies $\{n_1, n_2, ..., n_S\}$ follow a multinomial distribution with the probabilities $\{p_1, p_2, ..., p_S\}$. Then the expected number of species in the local sample with m individuals is given by,

$$S_{ind}(m) = S - \sum_{i=1}^{S} (1 - p_i)^m \tag{17}$$

however, as a matter of fact, the true p_i are unknown, and its abundance in the reference pool is observed as n_i and the total species observed is S_{obs}, then an unbiased estimator of $S_{ind}(m)$ is given by [9],

$$\hat{S}_{ind}(m) = S_{obs} - \sum_{n_i>0} \binom{n-n_i}{m} \Big/ \binom{n}{m}. \tag{18}$$

Calculate $\hat{S}_{ind}(m)$ over a range of m, we could plot them together to generate the rarefaction curve.

REMARKS

Comparison of species diversity across different sampling sites is one of the major tasks in ecological studies. A proper diversity index for doing the comparison will be extremely important. Otherwise the results based on unsuitable diversity metrics might be biased or totally misleading. Rarefaction methods introduced in this chapter would be of great helps in the comparison of species diversity across sites with various species number and abundance.

REFERENCES

[1] C. Shannon, "A mathematical theory of communications", *Bell Syst. Tech. J.* 27 379-423, 1948.

[2] J. Brower, J. Zar, C. von Ende, "Field and laboratory methods for general ecology", McGraw-Hill, Boston, 1998.

[3] E. Simpson, "The measurement of diversity", *Nature*. 163, 688, 1949.

[4] A. Chao, "Non-parametric estimation of the number of classes in a population", *Scand. J. Stat.* 11, 265-270, 1984.

[5] A. Chao, "Estimating the population size for capture-recapture data with unequal catchability", *Biometrics*. 43, 783-791, 1987.

[6] K. Burnham, W. Overton, "Estimation of the size of a closed population when capture probabilities vary among animals", *Biometrika*. 65, 625-633, 1978.

[7] E. Smith, G. van Belle, "Nonparametric estimation of species richness", *Biometrics*. 40, 119-129, 1984.

[8] N. Gotelli, A. Chao, "Measuring and estimating species richness, species diversity, and biotic similarity from sampling data", In: *Encycl. Biodivers.*, Academic Press, Waltham, MA, 2013: pp. 195-211.

[9] S. Hurlbert, "The nonconcept of species diversity: a critique and alternative parameters", *Ecology*. 52, 577-586, 1971.

CHAPTER 2

Functional Diversity

Abstract: This chapter provides some metrics for measuring functional diversity, another important diversity component in ecological studies. The metrics for functional diversity covered only some common used ones in this book, like functional richness, functional evenness, Rao's quadratic entropy index, functional divergence, functional regularity and functional attribute diversity. The computation for functional diversity of traits presented in categorical variables is also discussed.

Keywords: Categorical variable *versus* continuous variable, ecological indicators, ecosystem functioning, extinction risk, functional diversity, functional redundancy, functional richness and evenness, functional traits, molecular information, phenotype, statistical ecology.

INTRODUCTION

Species could not survive without functional traits. Functional traits are measured at individual level, thus quantifying the variation and diversity levels of morphological, physiological, behavioral, genetic and other biological aspects of species in terms of its individuals distributed over different spatial areas and temporal scales. Thus, functional traits could help reveal the evolutionary history and potential pathways of the species and predict its adaptation and responses to climate change. At many perspectives, functional diversity, measured from functional traits, become one of the most important biodiversity components in contemporary ecology and biodiversity analysis.

Here we introduce some functional diversity metrics but actually there are more than the ones that we can offer here. These metrics have been well interpreted in the user manual (https://sites.google.com/site/functionaldiversity/downloads) of a free software for computing functional diversity: FDiversity [1]. Thus, readers are encouraged to use the FDiversity to compute these relevant functional diversity indices to have direct feelings about their differences and meanings.

Different Functional Diversity Indices

Originality of Functional Diversity

The originality of functional diversity for a specific species is the distance between the position of that species in the diversity space and the mean position

Youhua Chen

over all the species [2-4]. The higher the value, the higher difference between the trait/environment diversity of the species and the hypothetical species with average trait/environment values is.

Uniqueness of Functional Diversity

The definition of uniqueness of functional diversity for each species is measured by the distance to the nearest neighbor in the functional space [4]. When the index is high, the species would thus inhabit areas with unique environmental characteristics when compared to other species (indicating low redundancy or high irreplaceability).

Quadratic Entropy

This index was proposed by Rao [5].

$$Q=\sum_{i=1}^{s-1}\sum_{j>1}^{s} d_{ij}p_i p_j = \frac{1}{2}p^T Dp \tag{1}$$

where s is the number of species, d_{ij} is the functional distance (can be the Euclidean distance among the traits) between species i and j. D is the distance matrix with elements d_{ij}. p is the vector of the relative abundance of species with elements p_i.

An unbiased estimator of Q could be [6],

$$\hat{Q}=2\sum_{i>j}^{s} d_{ij}\frac{n_i n_j}{n(n-1)} \tag{2}$$

where n_i is the number of individuals of species i, n is the total number of individuals across all the species. The variance of $\hat{Q}$ is given by,

$$\mathrm{var}(\hat{Q}) = \frac{4}{s(s-1)}\{(3-2s)(2\sum_{i>j}^{s} d_{ij}\frac{n_i n_j}{n^2})^2 + (s-2)\sum_{i,j,k}^{s} d_{ij}d_{ik}\frac{n_i n_j n_k}{n^2} + \sum_{i>j}^{s} d_{ij}\frac{n_i n_j}{n^2}\} \tag{3}$$

The variance of $\hat{Q}$ is said to be useful when large samples are tested.

By taking partial derivative of $\hat{Q}$ on the number of individuals of species j, one can get,

$$\begin{aligned}\partial Q / \partial n_j &= \partial\{2\sum_{i>j}^{s} d_{ij} \frac{n_i n_j}{n(n-1)}\} / \partial n_j \\ &\approx \partial\{2\sum_{i>j}^{s} d_{ij} \frac{n_i n_j}{n \times n}\} / \partial n_j \\ &= 2\sum_{i>j}^{s} d_{ij} \frac{n_i}{n^2} - 2\sum_{i>j}^{s} d_{ij} \frac{n_i n_j}{n^3} 2n \\ &= \frac{2}{n}(\sum_{i>j}^{s} d_{ij} \frac{n_i}{n} - \sum_{i>j}^{s} d_{ij} \frac{n_i n_j}{n^2}) \\ &= \frac{2}{n}(\sum_{i>j}^{s} d_{ij} \frac{n_i}{n} - Q)\end{aligned} \tag{4}$$

The above equation actually measured the relative contribution of species j to the total functional diversity. Because the first term of the last equality indicates the functional distance of species j to other species, while the second term is the mean functional divergence among all the species.

Functional Divergence [7]

Functional divergence index measures the variance of the functional traits of species that occur at a site. Its equation is given by,

$$FD = \frac{2}{\pi}\arctan(5V) \tag{5}$$

where V is the weighted variance of a trait x, expressed as,

$$V = \sum_{i=1}^{s} p_i (\ln x_i - \rho_x)^2 \tag{6}$$

where p_i is the relative abundance of species i, ρ_x is the weighted mean, being equal to $\rho_x = \sum_{i=1}^{s} p_i \ln x_i$.

Functional Regularity [8]

For a single trait, the species are ranked by increasing value of the trait values and the corresponding weighted difference of two consecutive species is,

$$EW_{i,i+1} = \frac{|x_{i+1} - x_i|}{p_{i+1} + p_i} \tag{7}$$

and the percent of the weighted difference is,

$$PEW_{i,i+1} = \frac{EW_{i,i+1}}{\sum_{i=1}^{s-1} EW_{i,i+1}} \tag{8}$$

Then the functional regularity index (FRO) is to choose the minimum between the consecutive pair's percent of weighted difference and the equal-probable space 1/(s-1) for summation as follows:

$$FRO = \sum_{i=1}^{s-1} \min(PEW_{i,i+1}, 1/(s-1)) \tag{9}$$

The above functional regularity index is for a single trait, for multiple trait situations, one can compute principal component analysis on the *m* traits and then extract the *m* principal components. Each principal component is analyzed using the above FRO index and also the standard deviation. Then, the equation for multiple-trait functional regularity index is given by,

$$MFRO = \sum_{i=1}^{m} SD_i \times FRO_i \tag{10}$$

Functional Attribute Diversity [9]

Functional attribute diversity is simply to sum up all the Euclidean distances across all the pairs of species based on their trait space.

The distance of two species *i* and *j* based on their trait values is simple the Euclidean distance of the traits as,

$$ED_{i,j} = \sqrt{\sum_{t=1}^{m} (x_{tj} - x_{ti})^2} \tag{11}$$

where *m* is the total number of traits.

Then the functional attribute diversity is calculated as,

$$FAD = \sum_{i=1}^{s-1} \sum_{j>i}^{s} ED_{i,j} \tag{12}$$

to compute functional attribute diversity across different samples, one can divide the above equation by the total species number for each sample as,

$$FAD_{sample} = \frac{FAD}{s(s-1)/2} \tag{13}$$

Functional Richness in One Dimension [10]

Functional richness (FR) for a single trait for a local sample is to measure the relative range of the trait values in the local sample in respect to the whole range of the trait across all the samples (*i.e.*, a whole community). The measurement of the range of the trait values in a local sample or the whole community is simply to calculate the absolute value between the largest trait value and the smallest trait value for the individuals or species found in the local sample or the whole community. In mathematical form, FR is computed as,

$$FR(i) = \frac{TR_i}{TR_{alls}} \tag{14}$$

where TR_i denotes the trait value range for species/individuals found in the local sample i; while TR_{alls} denotes the trait value range for all the species found in the whole community which is constituted by all the local samples.

If there are multiple traits, a simple method to measure FR is to take the average of FR for different traits to represent average FR.

Functional Evenness (FE) in One Dimension [10, 11]

FE is to measure whether the mean of the trait across species are distributed regularly within the occupied trait space. For example, if the mean trait value of species (the average of the trait values for the individuals of the species found in the sample) is distributed in the two-dimension Euclidean space evenly, that is, these trait values have equal distances between them and the corresponding species have equal abundances (individuals), then FE should be very high.

For a local sample i where there is $|S_c|$ species in the species set S_c, then for the trait space, there should be $|S_c|-1$ intervals. The computation of FE for a single trait is given by,

$$FE(i)=\sum_{j=1}^{|Sc|-1}\min\left\{\frac{1}{|Sc|-1},\frac{|x_{j+1}-x_j|/|a_{j+1}+a_j|}{\sum_{k=1}^{Sc|-1}|x_{k+1}-x_k|/|a_{k+1}+a_k|}\right\} \tag{15}$$

where x_j denotes the mean trait value for species *j* if it has multiple individuals in the local sample; the corresponding abundance is the count of the individuals and recorded as a_j. Please note here the trait values for different species from 1 to $|S_c|$ have been sorted from high to low. That is, $x_1 \geq x_2 \geq ... \geq x_{|S_c|}$.

For the local sample in which incidence data are collected (that is, no abundance information but only the presence/absence information is available for the species), then the corresponding FE should be calculated by ignoring the abundance information, that is,

$$FE(i)=\sum_{j=1}^{|Sc|-1}\min\left\{\frac{1}{|Sc|-1},\frac{|x_{j+1}-x_j|}{\sum_{k=1}^{Sc|-1}|x_{k+1}-x_k|}\right\} \tag{16}$$

If the trait is presented in categorical variable (*i.e.*, 1, 2, 3…), then the corresponding FE is given by,

$$FE(i)=\sum_{j=1}^{TN}\min\left\{\frac{a_j^1}{a^1},\frac{1}{TN}\right\} \tag{17}$$

where a_j^1 is the summed abundance for the species which has the categorical trait value *j*; a^1 is the total abundance across all the species in the sample (*i.e.*, $a^1=\sum_{j=1}^{TN}a_j^1=\sum_{j=1}^{|S_c|}a_j$); TN is the number of total categorical levels for the trait in the sample.

Similarly, if incidence data is available only, then FE for categorical trait variable is given by,

$$FE(i)=\sum_{j=1}^{TN}\min\left\{\frac{S_j}{|S_c|},\frac{1}{TN}\right\} \tag{18}$$

where S_j is the number of species that that have the trait categorical value j.

Functional Evenness (FE) in Multiple Dimensions [10, 11]

For a local sample *i*, the computation of FE for multiple trait situations is a bit complex, in which a minimal spanning tree (MST) should be constructed for connecting all the species that are found in the sample *i* based on the Euclidean/Gower distances between them. The Euclidean distance between a pair of species is constructed by using the values of the traits as the input. When the trait values are presented in discrete/categorical variables, then the Gower distance should be utilized.

After MST is constructed, the edges in the MST would be utilized to compute FE accordingly, which follows [11, 12],

$$FE(i) = \frac{\sum_{e \in E} \min\left\{ \frac{1}{|Sc|-1}, \frac{|e|/(a_e/a)}{\sum_{e' \in E} |e'|/(a_{e'}/a)} \right\} - \frac{1}{|Sc|-1}}{1 - \frac{1}{|Sc|-1}} \tag{19}$$

where E denotes the set of the edges in the MST; a_e is the summed abundance of the two species that is connected by the edge e; $|e|$ is the length of the edge e.

Functional Distance (FDIS) [13]

Computation of FDIS could be dependent on the multivariate statistics, depending on the researchers' own choice. Assuming for a local sample *i* where there is $|S_c|$ species in the species set S_c, it is required to generate a species-trait matrix in which element is the trait value for the corresponding species.

If one decides to reduce the dimensions of the trait space, then, a principal component analysis (PCA), principal coordinate analysis (PCoA) or multidimensional scaling (MDS) could be applied to the species-trait matrix to reduce its dimensions into two or three dimensions. Then, the resultant species score matrix (in two or three dimensions) is used to compute the weighted centroid of the reduced trait space.

If one doesn't want to reduce the trait space, the original species-trait matrix can be used directly to compute the centroid of the trait space.

The centroid of the trait space (for either the full or reduce space, dependent on whether the multivariate statistic is applied) in the sample i is weighted by the abundance of the species in the sample, which reads,

$$c(i) = [c_j] = \frac{\sum_{k=1}^{|Sc|} a_k x_{kj}}{\sum_{k=1}^{|Sc|} a_k} \tag{20}$$

where a_k is the abundance for the species k in the sample i; $c(i)$ is the weighted centroid vector across all the species in the trait space; c_j denotes the centroid value for the trait j in the trait space.

So now the functional distance (FDIS) for the local sample i is computed as,

$$FDIS(i) = \frac{\sum_{k=1}^{|Sc|} a_k d_k}{\sum_{k=1}^{|Sc|} a_k} \tag{21}$$

where d_k is the Euclidean distance from the position of the species k in the trait space to the weighted centroid c as above. In mathematical format, it is given by,

$$d_k = \sqrt{\sum_{t=1}^{TN} (c_t - s_{kt})^2} \tag{22}$$

where TN is the total number of traits for the full species-trait matrix or the total number of reduced dimensions; s_{kt} is the value for species k for the specific trait t in the full species-trait matrix or the axis t in the reduced trait space.

REMARKS

Based on the multi-dimensional volume definition of Hutchison's niche, functional diversity can be defined as the multi-dimensional volume of species' trait dispersion. The one-to-one mapping between niche and functional diversity suggested their close associations.

REFERENCES

[1] F. Casanoves, L. Pla, J. Rienzo, S. Diaz, FDiversity: "a software package for the integrated analysis of functional diversity", *Methods Ecol. Evol.* 2, 233-237, 2011.

[2] S. Villeger, J. Ramos Miranda, D. Flores Hernandes, D. Mouillot, "Contrasting changes in taxonomic *vs*. functional diversity of tropical fish communities after habitat degradation", *Ecol. Appl.* 20, 1512-1522, 2010.

[3] D. Bellwood, P. Wainwright, C. Fulton, A. Hoey, "Functional versatility supports coral reef biodiversity", *Proc. R. Soc. B Biol. Sci.* 273,101-107, 2006.

[4] L. Buisson, G. Grenouillet, S. Villeger, J. Canal, P. Laffaille, "Toward a loss of functional diversity in steram fish assemblages under climate change", *Glob. Change Biol.* 19, 387-400, 2013.

[5] C. Rao, "Diversity and dissimilarity coefficients-a unified approach", *Theor. Popul. Biol.* 21, 24-43, 1982.

[6] K. Shimatani, "On the measurement of species diversity incorporating species difference", *Oikos*. 93, 135-147, 2001.

[7] N. Mason, K. MacGillivrary, J. Steel, J. Wilson, "An index of functional diversity", *J. Veg. Sci.* 14, 571-578, 2003.

[8] D. Mouillot, W. Mason, O. Dumay, J. Wilson, "Functional regularity: a neglected aspect of functional diversity", *Oecologica*. 142, 353-559, 2005.

[9] B. Walker, J. Langridge, "Measuring functional diversity in plant communities with mixed life forms: a problem of hard and soft attributes", *Ecosystems*. 5, 529-538, 2002.

[10] N. Mason, D. Mouillot, W. Lee, J. Wilson, "Functional richness, functional evenness and functional divergence: the primary components of functional diversity", *Oikos*. 111,112-118, 2005.

[11] D. Schleuter, M. Daufresne, F. Massol, C. Argillier, "A user's guide to functional diversity indices", *Ecol. Monogr*. 80, 469-484, 2010.

[12] S. Villeger, N. Mason, D. Mouillot, "New multidimensional functional diveristy indices for a multifacted framework in functional ecology", *Ecology*. 89, 2290-2301, 2008.

[13] E. Laliberte, P. Legendre, "A distance-based framework for measuring functional diversity from multiple traits", *Ecology*. 91, 299-305, 2010.

Chapter 3: Phylogenetic Diversity

CHAPTER 3

Phylogenetic Diversity

Abstract: This chapter provides some metrics for measuring phylogenetic diversity, one of the most important diversity components in ecological studies. Phylogenetic diversity nowadays is widely used to evaluate the role of evolutionary history and evolutionary interaction among species on structuring contemporary biodiversity patterns and ecological community structure. The metrics for phylogenetic diversity could be divided into two categories: node-based and branch length-based metrics. For node-based indices, I and W give more weights on the sub-clades with fewer taxa. For branch-based metrics, the following metrics are introduced: total phylogenetic diversity, pendant edge, taxonomic distinctiveness, evolutionary distinctiveness, phylogenetic endemism, phylogenetic ancestral range index, and imperiled phylogenetic diversity index.

Keywords: Ancestor, Brownian motion of evolution, descendants, ecological indicators, evolutionary ecology, evolutionary history, external tips, extinction risk, molecular information, phylogenetic community structure, phylogenetic diversity, phylogenetic theory, statistical ecology, Tree of Life.

INTRODUCTION

Mapping and identifying the hotspots of the spatial biodiversity patterns and identification of species that have the most important conservation values is one of the major tasks in conservation biology [1]. In the study of biodiversity conservation, two important aspects [2] are to identify priority areas and species for conservation.

In previous studies, all species are treated equally without differentiation when conducting biodiversity analyses and statistics, which may be inadequate because such a treatment overlooks the information of species' functional differentiation and evolutionary history [3-8]. Therefore, to better quantify the conservation priorities of species and areas, integration of other information, for example, the evolutionary history of species, expressed as the phylogenetic relationships of the species, are crucial [7, 9-14]. In recent years, the estimation of species extinction in conservation biology has effectively incorporated the information of evolutionary history of species [7-9, 11, 15, 16].

MEASURE OF PHYLOGENETIC DIVERSITY BY NODE-BASED METHODS

One way to conduct node-based phylogenetic diversity (PD) calculation can be found in Posadas *et al.* [9]. In their work, two basic indices (I and W) should be

Youhua Chen

computed. In detail, index I is to assign a value of 1 to each tip species that belongs to a pair of terminal sister species. The most recent ancestor (one of the internal nodes) that includes only these two sister species will be assigned a value of 2. Each internal ancestor will receive a value being equal to the number of externally living descendants it has. Thus, index I refers to the number of phylogenetic groups to which a taxon belongs [9].

Comparatively, the PD index W measures the proportion that each species contributes to the total diversity of the clade. Specifically, index W will assign a value (i) to each tip species. This value is calculated as the number of internal nodes (ancestors) that connects the external tip species to the root of the tree. A weight (Q) was calculated as follows,

$$Q_j = \sum i / i_j \tag{1}$$

where j is equal to each specific taxon in the cladogram. The Q value for each taxon refers to the proportion of total diversity of the group that is contributed by this taxon. The PD measure (W) was calculated then:

$$W = Q_j / Q_{\min} \tag{2}$$

where $Q_{\min}$ refers to the lowest Q-value for the entire group.

MEASURE OF PHYLOGENETIC DIVERSITY BY BRANCH-BASED METHODS

Total Phylogenetic Diversity

Total phylogenetic diversity is simply to sum up all the branch lengths in a tree.

Pendant Edge

Pendant edge is a measure of a species' age since its speciation and calculated as the distance from any tip on a tree to where it subtends the tree of life [17].

Taxonomic Distinctiveness

The taxonomic distinctiveness Δ^+ index [18] is utilized since it is suitable for presence/absence data, which reads,

$$\Delta^+ = \{\sum\sum_{i<j} \omega_{ij}\} / \{s(s-1)/2\} \tag{3}$$

Here, s is the number of species present in the taxonomic/phylogenetic tree, ω_{ij} is the distinctness weight given to the path length linking species i and j in the classification tree. One simple weight is that ω_{ij} is equal to the number of internal nodes for linking the two species. However, if ω_{ij} is identical across different species, taxonomic distinctiveness actually is a node-based method for evaluating phylogenetic diversity.

Evolutionary Distinctiveness

This measures the length of the species' terminal branch plus its species-weighted shares of ancestral branches.

Phylogenetic Endemism Index (PE)

PE is defined by considering two attributes: branch lengths and distributional ranges of external species [19, 20]. The calculation of phylogenetic endemism is illustrated as the following equation. For a given external (or tip) species,

$$PE_s = \sum_{i \in N(s)} l_i / R_i \tag{4}$$

Where $N(s)$ is the node set for species which formed a unique path from the root to the tip species s. l_i is the branch length led by node i, R_i is the union of distributional ranges of species which were the descendants from the node i. As such, PE is simply the summation of weighted branch lengths which linked the root to the tip species. The weighting is the range size of species.

Phylogenetic Ancestral Range Index

When discussing the above problem in PE calculation, it is suggested to use the ancestral range A_i to indicate the true evolutionary history of species in that range. Assuming there is an internal node i (or ancestor species i) has the branch l_i which is one in the node set $N(s)$ that connect the root to the focusing tip species s. I quantify the weight of phylogenetic ancestral endemism as l_i / A_i for that internal branch. As such, when I calculate through all the weights for the nodes inside the node set $N(s)$ and sum them together to obtain phylogenetic ancestral endemism (PAE) index for the focused species s. The formula is written as,

$$PAE_s = \sum_{i \in N(s)} l_i / A_i \tag{5}$$

where A_i is the ancestral range size constructed for the internal node. Thus, my PAE is not directly associated to the contemporary distribution of species, but only the ancestral range information involved.

To incorporate evolutionary distinctiveness index (ED) efficiently [16, 17, 21], the above equation (2) could be further extended to the form as follows,

$$PAE_s = \sum_{i \in N(s)} l_i / (S_i \times A_i) \tag{6}$$

where S_i denotes the number of external species descended from the concerned internal node *i*.

The above equation could be utilized to evaluate the conservation priority of each of the external species in the phylogeny. When the distributional range of species is assumed to be identical, the above equation (3) is degenerated into the ED index [16, 17, 21]. For evaluating the conservation importance of different areas, one could simply sum the PAE values for the species found in each of the focused areas. In the subsequent analyses, I will utilize equation (3) to measure species' conservation values.

When the distributional range sizes for external species are unknown and only the distributional grids/quadrates of species are presented, one could still utilize the above equation (3). However, for this time the reconstruction of ancestral ranges must be carried out on each of distributional grids for the species. As such, the range size of a species is simply the summation of distributional grids where the species occurs. The range size for each of its ancestors is analogous by summing the occurrence probabilities of the focused ancestor across all the distributional grids.

Imperiled Phylogenetic Diversity Index (IPD Index)

IPD index has been developed in a recent study [20]. Following the same notations in the previous work [20], suppose a tree with *E* branch lengths, each branch *e* has a length *l(e)*, then the expected total loss of PD from the tree is,

$$IPD = \sum_{e \in E} (l(e) \prod_{x \in C(e)} p(ext)_x) \tag{7}$$

where $p(ext)_x$ is the assigned extinction risk probability of the external species *x* in the tree. $C(e)$ is the set of living descendants from the branch *e*.

Similar to the previous study [20], the IUCN threatened categories of species are defined and assigned specific extinction probabilities as follows, Least concern (0.0001), Near threatened (0.01), Vulnerable (0.1), Endangered (0.667), and Critically endangered (0.999).

REMARKS

Recent studies applied phylogenetic endemism index to search for centers of endemism. These endemic areas are full of endemic species or phylogenetically distinct species and thus deserve future conservation attention.

REFERENCES

[1] J. Diniz-Filho, L. Bini, M. Pinto, T. Rangel, P. Carvalho, S. Vieira, *et al.* "Conservation biogeography of anurans in Brazilian Cerrado", *Biodivers. Conserv.* 16, 997-1008, 2007.

[2] Y. Chen, "Prioritizing avian conservation areas in China by hotspot scoring, heuristics and optimisation", *Acta Ornithol.* 42, 119-128, 2007.

[3] R. Vane-Wright, C. Humphries, P. Williams, "What to protect? Systematics and the agony of choice", *Biol. Conserv.* 55, 235-254, 1991.

[4] D. Faith, "Conservation evaluation and phylogenetic diversity", *Biol. Conserv.* 61, 1-10, 1992.

[5] D. Faith, Quantifying biodiversity: "a phylogenetic perspective", *Conserv. Biol.* 16, 248-252, 2002.

[6] J. Diniz-Filho, "Phylogenetic autocorrelation analysis of extinction risks and the loss of evolutionary history in Felidae (Carnivora: Mammalia)", *Evol. Ecol.* 18, 273-282, 2004.

[7] A. Soutullo, S. Dodsworth, S. Heard, A. Mooers, "Distribution of correlates of carnivore phylogenetic diversity", *Anim. Conserv.* 8, 249-258, 2005.

[8] D. Fraser, L. Bernatchez, "Adaptive evolutionary conservation: towards a unified concept for defining conservation units", *Mol. Ecol.* 10, 2741-2752, 2001.

[9] P. Posadas, D. Esquivel, J. Crisci, "Using phylogenetic diversity measures to set priorities in conservation: an example from southern South America", *Conserv. Biol.* 15, 1325-1334, 2001.

[10] G. Barker, "Phylogenetic diversity: a quantitative framework for measurement of priority and achievement in biodiversity conservation", *Biol. J. Linn. Soc.* 76, 165-194, 2002.

[11] W. Sechrest, T. Brooks, G. da Fonseca, W. Konstant, R. Mittermeier, A. Purvis, *et al.*, "Hotspots and the conservation of evolutionary history", *Proc. Natl. Acad. Sci.* 99, 2067-2071, 2002.

[12] K. McGoogan, T. Kivell, M. Hutchison, H. Young, S. Blanchard, M. Keeth, *et al.*, "Phylogenetic diversity and the conservation biogeography of African primates", *J. Biogeogr.* 34, 1962-1974, 2007.

[13] M. Spathelf, T. Waite, "Will hotspots conserve extra primate and carnivore evolutionary history?", *Divers. Distrib.* 13, 746-751, 2007.

[14] F. Forest, R. Grenyer, M. Rouget, T.J. Davies, R.M. Cowling, D.P. Faith, *et al.* "Preserving the evolutionary potential of floras in biodiversity hotspots.", *Nature.* 445, 757-60. doi:10.1038/nature05587, 2007.

[15] S. Heard, A. Mooers, "Phylogenetically patterned speciation rates and extinction risks chance the loss of evolutionary history during extinctions", *Proc. R. Soc. Lodon Ser. B.* 267, 613-620, 2000.

[16] D. Redding, A. Mooers, "Incorporating evolutionary measures into conservation prioritization", *Conserv. Biol.* 20, 1670-1678, 2006.

[17] D. Redding, K. Hartmann, A. Mimoto, D. Bokal, M. Devos, A. Mooers, "Evolutionarily distinctive species often capture more phylogenetic diversity than expected", *J. Theor. Biol.* 251, 606-615, 2008.

[18] K. Clarke, R. Warwick, "A taxonomic distinctness index and its statistical properties", *J. Appl. Ecol.* 35, 523-531, 1998.

[19] D. Rosauer, S. Laffan, M. Crisp, S. Donnellan, L. Cook, "Phylogenetic endemism: a new approach for identifying geographical concentrations of evolutionary history", *Mol. Ecol.* 18, 4061-4072, 2009.

[20] R. Gudde, J. Joy, A. Mooers, "Imperiled phylogenetic endemism of Malagasy lemuriformes", *Divers. Distrib.* 19, 665-675, 2013.

[21] I. Martyn, T. Kuhn, A. Mooers, V. Moulton, A. Spillner, "Computing evolutionary distinctiveness indices in large scale analysis", *Algorithms Mol. Biol.* 7, 6, 2012.

CHAPTER 4

Multiple-Site Beta Diversity Methods with an Introduction of R Package "MBI"

Abstract: In this chapter, different multiple-site beta diversity metrics (totally 21 metrics) were introduced and calculated by using a R package "MBI", which was developed by the author of the book. The calculation of these multiple-site beta diversity metrics was actually quite easy and straightforward, especially when using the R package "MBI". In this chapter of the book, I will also present and compare the empirical applications of the metrics on the publicly available 290 presence/absence real-world matrices which were collected by ecologists. It is expected that the readers of the book could easily understand and learn these multiple-site beta diversity indices through practicing their own data by using the "MBI" package.

Keywords: Beta diversity, Bonferroni correction, computational ecology, diversity partitioning, multiple-site comparison, R statistical computing environment, software development, variance of species composition.

INTRODUCTION

In recent years, there is a growing trend that people are now trying to quantify beta diversity at community (or multiple-site) level. For example, Baselga [1,2] partitioned beta diversity into turnover and nestedness components at community levels. Later, it was found that these two indices were too insensitive to species loss in the community [3-5]. In a recent study, beta diversity was suggested to partition into the components of replacement and richness differences [3,4] at the pairwise-site level. However, the multiple-site counterparts of these metrics were still lacking, which would be developed in the present study. Finally, WNODF index [6] has been advocated as an effective indicator to quantify nestedness of the community.

Programs for calculating beta-diversity has been widely developed and there are a series of computer programs available for calculating pairwise indices, for example, function "*vegdist*" in the R packages "*vegan*" [7]. However, there are just a few programs specifically for multiple-site indices calculation, and they are typically only able to calculate one or two particular multiple-site indices. For example, WNODF index was available in a FORTRAN program developed by the authors [6] and "*vegan*" package [7]. Recently, [8] developed an R package "*betapart*" for calculating the nestedness and turnover indices mentioned above.

Youhua Chen

Thus, there was no program that can put different multiple-site indices together and ecologists should feel hard to compare different multiple-site beta diversity indices simultaneously since some indices were not included in computer programs. Thus, it is very necessary to develop a general package specifically for multiple-site beta diversity calculations, thus allowing ecologists to compare various indices easily for their own data sets. Inspired from this issue, I tried to collect different multiple-site diversity indices together and make their calculation become available using R scientific computing platform [9].

In this chapter, we demonstrated an R package, called "MBI" (http://cran.r-project.org/web/packages/MBI/), for calculating different indices of beta diversity for the purpose of multiple-site comparison. All the previously proposed indices, with the addition of some more indices developed in the present study, were included (Table **1**). Currently, there were 21 indices available, in which 10 novel ones were developed in the present study. Among these new indices, 7 were developed on the basis of pairwise indices which calculated the index values for only two sites each time iteratively and then average the values across all the pairs of sites in the community (Table **1**). The other 3 ones, including multiple-site versions of replacement, richness difference and Lennon's richness indices, were developed as the analogues of Baselga's nestedness and turnover indices (Table **1**). The deduction of these indices followed the same steps as Baselga's indices [2].

Table 1: Summary of 21 multiple-site diversity indices currently available in *MBI* package. S_i is the number of species found in site i, while S_T total number of species in the community. b_{ij} and b_{ji} are the number of species exclusive to sites *i* and *j* respectively when compared by pairs. T is the site number in the community. k_{ij} denotes the number of cells with lower values in column c_j compared to those in column c_i and N_j is the total number of non-empty cells in column c_j. k'_{ij} denotes the number of cells with lower values in row r_j compared to that for the row r_i and N'_j is the total number of non-empty cells in row r_j. N_{pairs} denotes the number of pairs of the sites

Index	Formula	Function in the MBI package	Reference
Baselga's full index	$\frac{\sum_{i<j}\max(b_{ij},b_{ji})+\sum_{i<j}\min(b_{ij},b_{ji})}{2(\sum_i S_i - S_T)+(\sum_{i<j}\min(b_{ij},b_{ji}))+(\sum_{i<j}\max(b_{ij},b_{ji}))}$	*cfull(data)*	[1]

Table 1: contd….

Baselga's nestedness	$\frac{\sum_{i<j}\max(b_{ij},b_{ji})-\sum_{i<j}\min(b_{ij},b_{ji})}{2(\sum_i S_i - S_T)+\sum_{i<j}\max(b_{ij},b_{ji})+\sum_{i<j}\min(b_{ij},b_{ji})} \times \frac{(\sum_i S_i - S_T)}{(\sum_i S_i - S_T)+\sum_{i<j}\min(b_{ij},b_{ji})}$	*cn(data)*	[1,2]
Baselga's turnover	$\frac{\sum_{i<j}\min(b_{ij},b_{ji})}{(\sum_i S_i - S_T)+(\sum_{i<j}\min(b_{ij},b_{ji}))}$	*ct(data)*	[1,2]
Replacement	$2\times\frac{\sum_{i<j}\min(b_{ij},b_{ji})}{\left(\sum_i S_i - S_T\right)+\sum_{i<j}(b_{ij}+b_{ji})}$	*crep(data)*	This work; [4]
Richness difference	$\frac{\sum_{i<j}\max(b_{ij},b_{ji})-\sum_{i<j}\min(b_{ij},b_{ji})}{2(\sum_i S_i - S_T)+\sum_{i<j}\max(b_{ij},b_{ji})+\sum_{i<j}\min(b_{ij},b_{ji})}$	*crich(data)*	This work; [4]
WNODF	$WNODF = \frac{2(WNODFc+WNODFr)}{m(m-1)+n(n-1)}$ (overall index) $WNODFr = 100\sum_{i=1}^{m-1}\sum_{j=i+1}^{m}\frac{k'_{ij}}{N'_j}$ (index for rows) $WNODFc = 100\sum_{i=1}^{n-1}\sum_{j=i+1}^{n}\frac{k_{ij}}{N_j}$ (index for columns)	*wnodf(data)*	[3,6,10]
Whittaker's beta	$\frac{S_T}{\sum_i S_i / T}$	*wbeta(data)*	[11]

Table 1: contd….

Harrison's dissimilarity	$(\frac{S_T}{\sum_i S_i / T} - 1)/(T-1)$	*harrison(data)*	[12]
Diserud-Odegaard's index	$1-(\frac{S_T}{\sum_i S_i / T} - 1)/(T-1)$	*do(data)*	[13]
Harrison's turnover	$(\frac{S_T}{\max(S_i)} - 1)/(T-1)$	*ht(data)*	[12]
Williams's turnover	$1-\frac{\max(S_i)}{S_T}$	*wt(data)*	[14]
Lennon richness	$\frac{2(\sum_{i<j} \max(b_{ij}, b_{ji}) - \sum_{i<j} \min(b_{ij}, b_{ji}))}{2(\sum_i S_i - S_T) + \sum_{i<j} \max(b_{ij}, b_{ji}) + \sum_{i<j} \min(b_{ij}, b_{ji})}$	*cl(data)*	This work; [15]
Mean pairwise Jaccard distance	$\frac{1}{N_{pairs}} \sum_{i<j}^{N_{pairs}} \frac{b_{ij} + b_{ji}}{a + b_{ij} + b_{ji}}$	*mjaccard(data)*	This work
Mean pairwise Sorensen distance	$\frac{1}{N_{pairs}} \sum_{i<j}^{N_{pairs}} \frac{b_{ij} + b_{ji}}{2a + b_{ij} + b_{ji}}$	*msorensen(data)*	This work; [16]
Mean pairwise nestedness	$\frac{1}{N_{pairs}} \sum_{i<j}^{N_{pairs}} (\frac{\lvert b_{ij} - b_{ji} \rvert}{2a + b_{ij} + b_{ji}} \times \frac{a}{a + \min(b_{ij}, b_{ji})})$	*mn(data)*	This work; [2]
Mean pairwise turnover	$\frac{1}{N_{pairs}} \sum_{i<j}^{N_{pairs}} \frac{\min(b_{ij}, b_{ji})}{a + \min(b_{ij}, b_{ji})}$	*mt(data)*	This work; [2]

Table 1: contd….

Mean pairwise replacement	$\frac{2}{N_{pairs}}\sum_{i<j}^{N_{pairs}}\frac{\min(b_{ij},b_{ji})}{a+b_{ij}+b_{ji}}$	*mrep(data)*	This work; [4]
Mean pairwise richness difference	$\frac{1}{N_{pairs}}\sum_{i<j}^{N_{pairs}}\frac{\lvert b_{ij}-b_{ji}\rvert}{a+b_{ij}+b_{ji}}$	*mrich(data)*	This work; [4]
Mean pairwise Lennon richness	$\frac{2}{N_{pairs}}\sum_{i<j}^{N_{pairs}}\frac{\max(b_{ij},b_{ji})-\min(b_{ij},b_{ji})}{2a+\min(b_{ij},b_{ji})+\max(b_{ij},b_{ji})}$	*ml(data)*	This work; [15]

Thus, our multiple-site indices could be classified into two categories. The first group contained those deduced from mathematical formulation (called as "multiple-site indices without pairwise calculations") [1, 2], while the other group involved those with straightforward calculations by taking the averages over all the pairwise values (called as "mean pairwise indices"). The latter category included 7 new indices developed in the present study, while the former one was comprised of the other 14 indices (Table **1**).

Because we introduced a series of new indices for measuring multiple-site diversity, in addition to the demonstration of MRI package, another objective of the study is to compare the effectiveness of these new indices on quantifying multiple-site diversity patterns with reference to the available ones previously published.

Empirical Evaluation of the Multiple-Site Beta Diversity Indices

Here we tested the indices using the public 290 presence/absence real-world matrices, which were compiled by the previous work [17]. These real-world matrices have varying species and site numbers. Thus, they can help quantify the relationships of these multiple-site indices from an empirical perspective.

A Pearson's correlation analysis was performed to compare all the indices and the significance levels of the correlations were tested with Bonferroni correction when comparing the values for different indices across the 290 matrices.

RESULTS AND DISCUSSION

A Comparison of All Multiple-Site Indices

When observing the index values for the 290 real-world ecological matrices, most of the indices were tightly associated as indicated by Table **2**. Typically each pair from the 21 available metrics has high correlation values and significance level of P values<0.001.

Table 2: Pearson's correlation analysis and significance test among the multiple-site diversity indices across 290 real-world presence/absence matrices. Upper triangular zone indicated the P values, while lower triangular zone showed the correlation coefficients. The abbreviations of each index were adopted from the function names showed in Table **1**

	cn	ct	crep	crich	wnodfT	wnodfC	wnodfR	wbeta	harrison	ht	wt	do	cl	msorensen	mjaccard	mn	mt	mrep	mrich	ml	cfull
cn	0	0	0	0	0	0	0	0	0	0	0	0	0	0	0	0	0	0	0.03	0	0
ct	-0.76	0	0	0	0	0	0	0	0	0	0	0	0	0	0	0	0	0	0	0.14	0
crep	-0.87	0.77	0	0	0	0	0	0	0	0	0	0	0	0	0	0	0	0	0.18	0	0
crich	0.69	-0.17	-0.72	0	0	0	0	0	0	0	0	0	0	0.42	0.54	0	0	0	0	0	0
wnodfT	0.56	-0.43	-0.56	0.46	0	0	0	0	0	0	0	0	0	0	0	0	0	0	0.27	0	0.01
wnodfC	0.55	-0.43	-0.54	0.43	0.93	0	0	0	0	0	0	0	0	0	0	0	0	0	0.11	0	0
wnodfR	0.59	-0.56	-0.59	0.37	0.48	0.34	0	0	0	0	0	0	0	0	0	0	0	0	0.14	0	0
wbeta	-0.31	0.61	0.22	0.23	-0.27	-0.29	-0.26	0	0	0.53	0	0	0	0	0	0.39	0	0	0	0	0
harrison	-0.24	0.26	0.39	-0.27	-0.42	-0.43	-0.27	0.18	0	0	0	0	0	0	0	0	0	0	0	0.08	0
ht	-0.5	0.39	0.68	-0.59	-0.48	-0.46	-0.4	0.03	0.8	0	0	0	0	0	0	0	0	0	0.08	0	0.03
wt	-0.78	0.8	0.85	-0.48	-0.62	-0.62	-0.55	0.5	0.48	0.63	0	0	0	0	0	0	0	0	0	0	0
do	0.24	-0.26	-0.39	0.27	0.42	0.43	0.27	-0.18	-0.99	-0.8	-0.48	0	0	0	0	0	0	0	0	0.08	0
cl	0.69	-0.17	-0.72	1	0.46	0.43	0.37	0.23	-0.27	-0.59	-0.48	0.27	0	0.42	0.54	0	0	0	0	0	0
msorensen	-0.34	0.67	0.46	0.04	-0.35	-0.37	-0.4	0.61	0.68	0.5	0.62	-0.68	0.04	0	0	0.1	0	0	0	0	0
mjaccard	-0.35	0.68	0.48	0.03	-0.37	-0.38	-0.38	0.58	0.68	0.5	0.63	-0.68	0.03	0.98	0	0.16	0	0	0	0	0
mn	0.84	-0.48	-0.82	0.85	0.48	0.44	0.51	-0.05	-0.21	-0.54	-0.61	0.21	0.85	-0.09	-0.08	0	0	0	0	0	0.43
mt	-0.71	0.79	0.79	-0.39	-0.53	-0.52	-0.58	0.52	0.66	0.69	0.82	-0.66	-0.39	0.86	0.84	-0.59	0	0	0	0.1	0
mrep	-0.8	0.76	0.93	-0.63	-0.56	-0.54	-0.6	0.33	0.59	0.77	0.83	-0.59	-0.63	0.67	0.68	-0.73	0.92	0	0	0	0
mrich	0.12	0.33	-0.07	0.51	-0.06	-0.09	-0.08	0.57	0.48	0.1	0.21	-0.48	0.51	0.82	0.79	0.37	0.47	0.16	0	0	0
ml	0.55	-0.08	-0.57	0.84	0.24	0.2	0.26	0.33	0.1	-0.34	-0.24	-0.1	0.84	0.39	0.38	0.81	-0.09	-0.39	0.79	0	0
cfull	-0.24	0.8	0.35	0.37	-0.13	-0.15	-0.3	0.64	0.16	0.12	0.49	-0.16	0.37	0.71	0.71	0.04	0.55	0.41	0.62	0.37	0

One index, the mean pairwise richness difference index, had low consistent degrees of measurements when correlated it to other indices. Thus, except for this index, other indices shared a high congruence across the 290 matrices principally. Therefore, most of the diversity indices should be appropriate for measuring and comparing beta diversity among multiple communities.

Effectiveness of mean pairwise indices as the surrogates of other multiple-site indices without pairwise calculations.

We found that many of the mean pairwise indices were tightly associated to the ones that donot take the average of pairwise calculations, as shown in Fig. **1**.

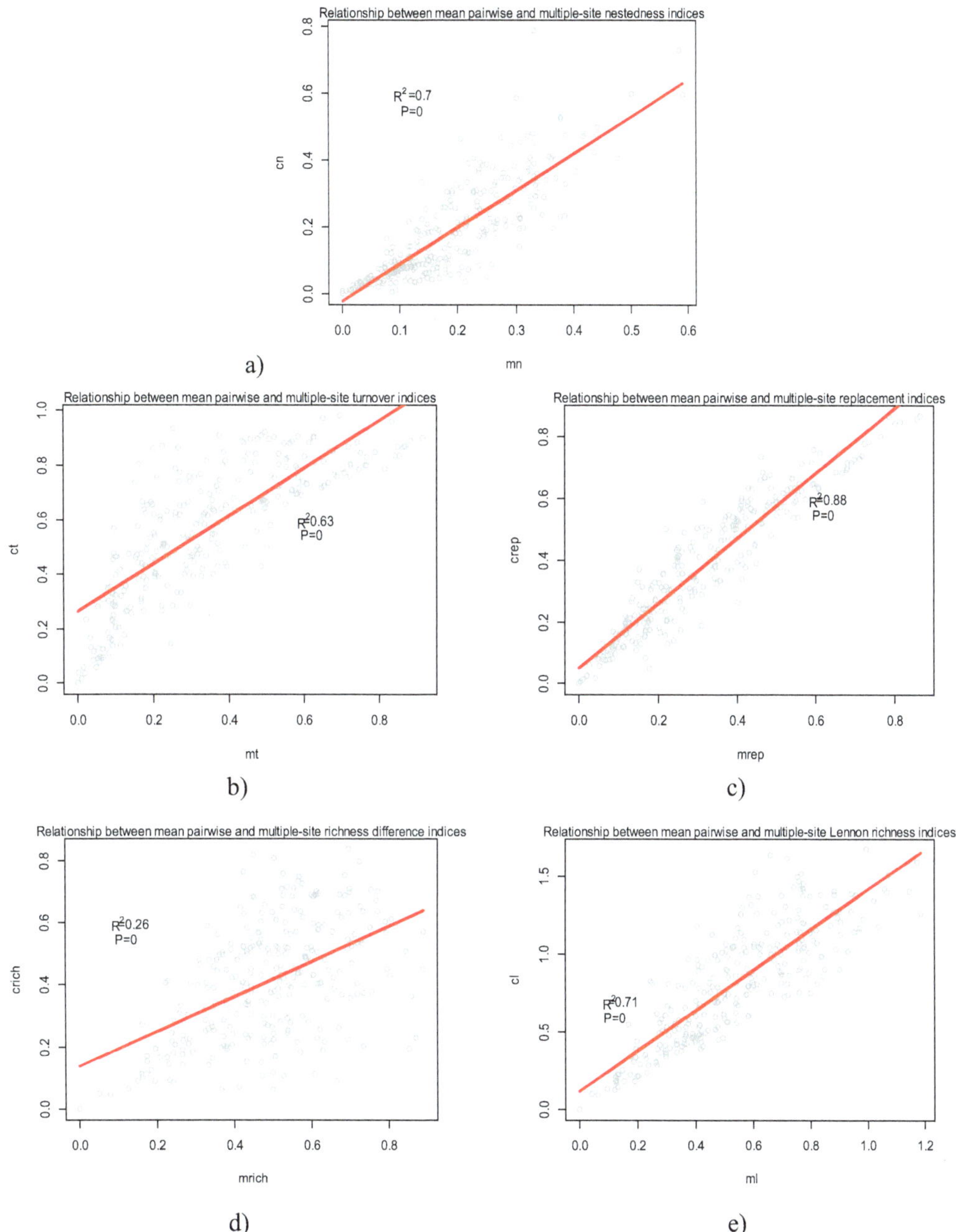

Figure 1: Effectiveness of using mean pairwise beta diversity indices as the surrogates of multiple-site indices. a) relationship between mean pairwise and multiple-site versions of nestedness indices; b) relationship between mean pairwise and multiple-site versions of turnover indices; c) relationship between mean pairwise and multiple-site versions of replacement indices; d) relationship between mean pairwise and multiple-site versions of richness difference indices; e) relationship between mean pairwise and multiple-site versions of Lennon's richness indices.

All the four multiple-site indices without pairwise calculations could be well represented by their counterparts of mean pairwise indices, including nestedness, turnover, Lennon's richness and replacement indices. It is worth noting that the richness difference multiple-site index without pairwise calculation was not strongly correlated to its counterpart (Fig. **1**). Thus, mean pairwise beta diversity indices proposed in the present work could be used to measure overall beta diversity for the whole community effectively in comparison to other multiple-site indices without iterative calculations of indices for each pair of sites.

REFERENCES

[1] Baselga, A. Jimenez-Valverde, G. Niccolini, "A multiple-site similarity measure independent of richness", *Biol. Lett.* 3, 642-645, 2007.

[2] A. Baselga, "Partitioning the turnover and nestedness components of beta diversity", *Glob. Ecol. Biogeogr.* 19, 134-143. doi:10.1111/j.1466-8238.2009.00490.x, 2010.

[3] M. Almeida-Neto, D. Frensel, W. Ulrich, "Rethinking the relationship between nestedness and beta diversity: a comment on Baselga (2010)", *Glob. Ecol. Biogeogr.* 21, 772-777, 2012.

[4] J. Carvalho, P. Cardoso, P. Gomes, "Determining the relative roles of species replacement and species richness differences in generating beta-diversity patterns", *Glob. Ecol. Biogeogr.* 21, 760-771, 2012.

[5] J. Carvalho, P. Cardoso, P. Borges, D. Schmera, J. Podani, "Measuring fractions of beta diversity and their relationships to nestedness: a theoretical and empirical comparison of novel approaches.", *Oikos.*, Doi: 10.1111/j.1600-0706.2012.20980.x. doi:10.1111/j.1600-0706.2012.20980.x, 2012.

[6] M. Almeida-Neto, W. Ulrich, "A straightforward computational approach for measuring nestedness using quantitative matrices", *Environ. Model. Softw.* 26, 173-178, 2011.

[7] J. Oksanen, F. Blanchet, R. Kindt, P. Legendre, P. Minchin, R. O'Hara, *et al.*, vegan: Community Ecology Package. R package version 2.0-4. http://CRAN.R-project.org/package=vegan, (2012).

[8] A. Baselga, C. Orme, "betapart: an R package for the study of beta diversity", *Methods Ecol. Evol.* 3, 808-812, 2012.

[9] R Development Core Team, R: A Language and Environment for Statistical Computing, Vienna, Austria. ISBN 3-900051-07-0, URL http://www.R-project.org., (2013).

[10] M. Almeida-Neto, P. Guimaraes, P. Guimaraes, R. Loyola, W. Ulrich, "A consistent metric for nestedness analysis in ecological systems: reconciling concept and measurement", *Oikos.* 117,1227-1239, 2008.

[11] R. Whittaker, "Vegetation of the Siskiyou Mountains, Oregon and California", *Ecol. Monogr.* 30, 280-338, 1960.

[12] S. Harrison, S. Ross, J. Lawton, "Beta-diversity on geographic gradients in Britain", *J. Anim. Ecol.* 61, 151-158, 1992.

[13] O. Diserud, F. Ødegaard, "A multiple-site similarity measure", *Biol. Lett.* 3 20-22, 2007.

[14] P. Williams, "Mapping variations in the strength and breadth of biogeographic transition zones using species turnover", *Proc. R. Soc. B Biol. Sci.* 263, 579-588, 1996.

[15] J. Lennon, P. Koleff, J. Greenwood, K. Gaston, "The geographical structure of British bird distributions: diversity, spatial turnover and scale", *J. Anim. Ecol.* 70, 966-979, 2001.

[16] T. Sorensen, "A method of establishing groups of equal amplitude in plant sociology based on similarity of species content, and its application to analyses of the vegetation on Danish commons", K. Dan. Vidensk. Selsk. Biolgiske Skr. 5, 1-34, 1948.

[17] W. Atmar, B. Patterson, The nestedness temperature calculator: a visual basic program, including 294 presence/absence matrices. AICS Research, Inc., University Park, NM and The Field Museum, Chicago, IL., (1995).

CHAPTER 5

Species-Site Compositional Matrix Comparison Methods

Abstract: Matrix comparison is an important element in community ecology because the distribution of different species over different areas is usually compiled in a form of a species-site matrix. Thus, if we have different species-site matrices (different species groups in the same sites or the same species assemblage in different non-spatially overlapped regions), we may want to know whether the resultant species-site matrices present similarities. In other words, we may want to compare the distributional congruence of the same set of species across different non-spatially overlapped regions or the distributional congruence of different groups of species in the same set of sites for a region. Thus, statistical methods for matrix comparison could help address these ecological questions. In this chapter, I will present the major statistical methods for conducting matrix similarity comparison. I will use the distribution of bird and mammal species in Hainan Island of China as the case study and example to demonstrate some of the matrix comparison methods introduced in the chapter.

Keywords: Avian biology, community ecology, compositional similarity, distributional congruence and incongruence, island biogeography, island ecology, mammalian biology, Mantel statistic, matrix operation and decomposition, multiscale analyses, partial correlation test, Procrustes statistics, spatial distribution and diversity, spatial ecology, species-site matrix.

INTRODUCTION

Distributional discordance and concordance patterns have been often revealed in macroecological studies [1]. However, most previous studies relied on the comparison of visual mapping of diversity patterns, which may be biased and inaccurate. Statistical methods on matrix comparison actually have been well developed [2]. The application of these statistical methods on the distributional matrices could help us reveal the distributional concordance patterns rigorously [3]. As such, the purpose of the present study is to employ Mantel test and Procrustes test to uncover the concordance patterns of spatial distribution of species. The methods introduced here could quantify the influence of spatial autocorrelation and environmental filtering on structuring species distribution patterns.

Mantel test has been widely used to compare the matrices [4-7]. However, its legitimacy has been under debate recently [4, 8, 9]. The questioning of Mantel test is due to the requirement of distance transformation, which breaks the original

Youhua Chen

variance structure of the matrix [3, 4]. Therefore, to maintain original covariance-variance structure of species-site matrix, it might be better to perform other matrix comparison methods without distance transformation. Procrustes test is one of such options [2], which utilized the original species-site matrices as the input. Procrustes analysis has been applied in some previous studies [3, 10, 11].

In this section, we introduced and compared both the above-mentioned methods for comparing matrix similarity. Distributional concordance of bird and mammal species in China was used as a case study. Moreover, multiscale-version analysis of both methods were introduced as well. Correspondingly, the Mantel test is better called as Mantel correlogram [5], while Procustes test as the Procustes correlogram. The significance of the test was computed using the permutation test by 1000 runs for each spatial distance class.

STATISTICAL METHODS FOR MATRIX COMPARISON

Mantel Test

We introduced the standardized Mantel statistic r_M to quantify matrix similarity [2, 6], For any two distance matrices *X* and *Y*,

$$r_M = \frac{1}{D-1}\sum_{i=1}^{n-1}\sum_{j=i+1}^{n}\left(\frac{x_{ij}-\overline{x}}{s_x}\right)\left(\frac{y_{ij}-\overline{y}}{s_y}\right) \tag{1}$$

where $D = n(n-1)/2$ is the number of distances in the upper triangular part of each matrix. $\overline{x}$ and $\overline{y}$ are the means of the values in each of the lower-triangular distance matrices *X(d)* and *Y(d)* for the given distance class *d*. s_x and s_y are their standard deviations.

Mantel Correlogram

For different distance-class matrices, the mantel test was applied individually, thus resulting into a series of Mantel statistics as a correlogram. The calculation of Mantel correlogram has the form as [5]. For any two lower-triangular (or upper-triangular) distance matrices *X* and *Y*, the Mantel's $r_M(d)$ correlogram index for a given distance class *d* is calculated as follows. Given that the lower-triangular distance matrices *X(d)* and *Y(d)* for the given distance class *d*,

$$r_M(d) = \frac{1}{W(d)-1}\sum_{i=2}^{n}\sum_{j=1}^{i-1} w_{ij}\left(\frac{x_{ij}-\overline{x}}{s_x}\right)\left(\frac{y_{ij}-\overline{y}}{s_y}\right) \tag{2}$$

where $\bar{x}$ and $\bar{y}$ are the means of the values in each of the lower-triangular (or upper-triangular) distance matrices *X(d)* and *Y(d)* for the given distance class *d*. s_x and s_y are their standard deviations. *W(d)* is the number of distances in the lower triangular part of each distance matrix. The weights w_{hi} have value 1 for pairs of sites belonging to the distance class *d* and 0 otherwise. *n* is the number of observations.

Partial Mantel Correlogram

The partial correlogram method was described and applied in some previous studies [12, 13]. The partial correlogram method is to obtain three sub-matrices *X(d)*, *Y(d)* and *Z(d)* based on the distance class *d* from the original matrices *X, Y* and *Z*, respectively. Here matrix *Z* (climate matrix in this example) is the one needed to be controlled for its effects on matrices *X* and *Y*. Then, the normal partial Mantel test was performed on the sub-matrices *X(d)*, *Y(d)* and *Z(d)* [2].

Multiscale Mantel Correlogram

For different distance-class matrices, the mantel test was applied individually, thus resulting into a series of Mantel statistics as a correlogram. The calculation of Mantel correlogram has the form as [5]. For any two distance matrices *X* and *Y*, the Mantel's $r_M(d)$ correlogram index for a given distance class *d* is calculated as follows. Given that the lower-triangular distance matrices *X(d)* and *Y(d)* for the given distance class *d*,

$$r_M(d) = \frac{1}{W(d)-1}\sum_{i=2}^{n}\sum_{j=1}^{i-1} w_{ij}\left(\frac{x_{ij}-\bar{x}}{s_x}\right)\left(\frac{y_{ij}-\bar{y}}{s_y}\right) \tag{3}$$

where $\bar{x}$ and $\bar{y}$ are the means of the values in each of the lower-triangular distance matrices *X(d)* and *Y(d)* for the given distance class *d*. s_x and s_y are their standard deviations. *W(d)* is the number of distances in the lower triangular part of each distance matrix. The weights w_{hi} have value 1 for pairs of sites belonging to the distance class *d* and 0 otherwise. *n* is the number of observations.

Procrustes Analysis

For two matrices which have the same row number (representing the sites), X and Y, we performed the singular value decomposition as [2, 14-16],

$$X^T Y = VWU^T \tag{4}$$

the Procrustes statistic is,

$$m_{12} = 1 - Tr(W)^2 \quad (5)$$

Partial Procrustes Correlogram

Differently, the partial Procrustes correlogram is to perform multivariate regression analysis of matrix *Z(d)* on the sub-matrices *X(d)* and *Y(d)*. Then, the residual matrices $X(d)_{res}$ and $Y(d)_{res}$ are used to perform normal Procrustes analysis following equations (5) and (6) [3, 16]. This method was run through across different distance classes to obtain the partial Procrustes correlogram.

Multiscale Procrustes Correlogram

The multiscale Procrustes correlogram could be constructed as follows. For two matrices with the given distance class *d*, *X(d)* and *Y(d)*, the Procrustes statistic $m(d)$ for the matrices for distance class *d* could be computed similarly as the above equations (2)-(3),

$$X(d)^T Y(d) = V_d W_d U_d^T \quad (6)$$

$$m_{12}(d) = 1 - Tr(W_d)^2 \quad (7)$$

To construct appropriate sub-matrices *X(d)* and *Y(d)* for distance class *d*, I utilized the following strategy. Firstly, the distances of sites are calculated. Then, for each pair of sites, if their distance is fallen into the distance class *d*, they will be kept in the sub matrices *X(d)* and *Y(d)*. For other sites, all the species' records were removed, resulting zero vectors for these sites. As such, the generated matrices *X(d)* and *Y(d)* for different distance classes were used to perform Procrustes test.

Test of Significance Using Permutation Techniques

The significance of statistics for Mantel statistic from randomness could be tested by permutation techniques [2]. For all the cases including global test without multiscale manipulation, multiscale non-partial and partial correlograms, 1000 permutations were run separately and the significance of probability was calculated individually. All the above statistical analyses could be done in R computing platform [17] using "*vegan*" package [18].

Test of Significance Using Fisher's Z-Transformation Method

Fisher's Z-transformation method doesn't require conducting permutation test, thus could extensively save the computational time and accelerate the computational speed to report results. Such a method thus is very applicable to the large matrices.

$$W = \frac{1}{2}\log(\frac{1+r_M}{1-r_M}) \tag{8}$$

Under the null hypothesis, the Mantel statistic r_M is not significantly different from $r_0 = 0$, as such, W followed an asymptotic normal distribution [19, 20]. Then, the expectation and variance of W statistic were given by [20],

$$E(W) = \frac{1}{2}\log(\frac{1+r_0}{1-r_0}) = 0 \tag{9}$$

$$Var(W) = \frac{1}{D-3} \tag{10}$$

it could be found that,

$$Z = \frac{W - E(W)}{\sqrt{Var(W)}} = \frac{W}{\sqrt{Var(W)}} = \sqrt{(D-3)} \times W \sim N(0,1) \tag{11}$$

which is directly from Theorem 3.2 in page 80 of the book [20].

For a unit normal distribution, the 2.5% and 97.5% quantiles were -1.96 and 1.96 respectively. As a consequence, when $Z \geq 1.96$ or $Z \leq -1.96$, we could ascertain that the observed Mantel statistic is significantly different from the expected random ones with $P \leq 0.05$. For the higher significant level $P \leq 0.001$, the corresponding Z values are $Z \geq 3.29$ (99.95% quantile) or $Z \leq -3.29$ (0.05 % quantile).

For small sample sizes (*e.g.*, $D \leq 25$), we could have a Hotelling's transform [20, 21] for the correction of sampling biases as,

$$W^* = W - \frac{3W + \tanh(W)}{4(D-1)} \tag{12}$$

here,

$$Var(W^*) = \frac{1}{D-1} \tag{13}$$

so, the corrected Z value for small data sets reads,

$$Z^* = \frac{W^* - E(W^*)}{\sqrt{Var(W^*)}} = \sqrt{(D-1)} \times W^* \sim N(0,1) \tag{14}$$

A PRACTICAL EXAMPLE USING BIRD AND MAMMAL DISTRIBUTION IN HAINAN ISLAND OF CHINA

An application on the distributional concordance of birds and mammals in Hainan Island (China).

Data Set

The presence-absence matrices for the distribution of birds and mammals were investigated. The species-site presence-absence data were collected form the previous studies [22, 23]. The corresponding twelve environmental variables including physical and climatic ones were collected from the previous studies as well [22, 23], which were annual mean temperature, annual mean precipitation, isothermality, altitude, mean monthly temperature range, temperature seasonality, maximum temperature of warmest month, minimum temperature of coldest month, temperature annual range, precipitation of wettest month, precipitation of driest month, and precipitation seasonality.

The following ecological hypotheses have been proposed for the test: (1) Bird and mammal distribution should have high congruence of distribution since there are a great amount of species in the region (156 birds and 76 mammals for 19 administrative sites, the locations were presented in Fig. **1**). Therefore, we hypothesized that across all the distance classes, the congruence of avian and mammalian distribution should be usually significant. (2) Given that the long-distance migration abilities of both mammals and birds, and because of the relative homogeneity of environmental condition across different sites in the island, we expect that environmental filtering plays a minor role in structuring species distribution. Therefore, we hypothesize that distributional concordance of birds and mammals showed a pattern of environmental independence.

Three distance numbers were considered for performing multiscale analysis, including N=5, 10, and 15 respectively. Accordingly, the distance interval for each distance number is determined as $\frac{\max(D_X)-\min(D_X)}{N}$, where D_X is Euclidean distance matrix constructed from the original matrix X. This distance interval is used to construct the sub-matrix $X(d)$.

RESULTS

Distributional Concordance of Birds and Mammals

Before correcting environmental dependence, the Mantel statistic r_M had the value of 0.4579 with $P<0.01$. After partialling out the influence of environment, the r_M statistic changed slightly with the value 0.4565 ($P<0.01$). As such, the second hypothesis was verified: environment hardly played a role on structuring distributional congruence patterns of birds and mammals in Hainan Island. Procrustes analysis gave similar results, before removing environmental forces, the statistic m_{12}= 0.1166 ($P<0.01$). After controlling environment, the statistic m_{12}=0.0785 ($P<0.01$).

Multiscale Distributional Concordance Using (Partial) Mantel Correlogram

When ignoring the influence of environment and without performing partial correlogram, we found that across different distance classes for distance number N=5 (Fig. **1a**), the distribution of birds and mammals has the significant patterns of correlation among 3 out of 5 distance classes. Thus, the first hypothesis was partially evidenced, and the distributional concordance of mammals and birds could vary according to different spatial scales. However, when more distance classes are considered (for distance numbers N=10 and 15), there are more distance classes being insignificant for the Mantel statistic (Figs. **1c** and **1e**). As such, the first hypothesis was less supported when more spatial scales were examined.

In light with our prediction, after the removal of environmental dependence by performing partial correlogram, the distributional concordance patterns changed slightly (*e.g.*, in distance number N=10) and the significance of the Mantel statistics was kept (Figs. **1b**, **1d** and **1f**). As such, the distributional congruence of birds and mammals in Hainan Island is not determined by the environmental conditions. Other local or regional influence should be dominant, for example, the immigration of species from mainland region.

Multiscale Distributional Concordance Using (Partial) Procrustes Correlogram

In either partial or non-partial situations (Fig. **2**), Procrustes statistics were always close to the global value (Procrustes statistic on the original bird and mammal data without any partitioning according to distance classes). Given the global Procrustes statistics were low as 0.1166 with significance, all the statistics across different distance classes have the similar results. As such, it was observed that all the m_{12} had the significance level with $P<0.01$. Therefore, different spatial scales didn't change the original matrix similarity between birds and mammals in Hainan Island. The matrix comparison was not influenced by spatial distance classes when one utilized Procrustes correlogram.

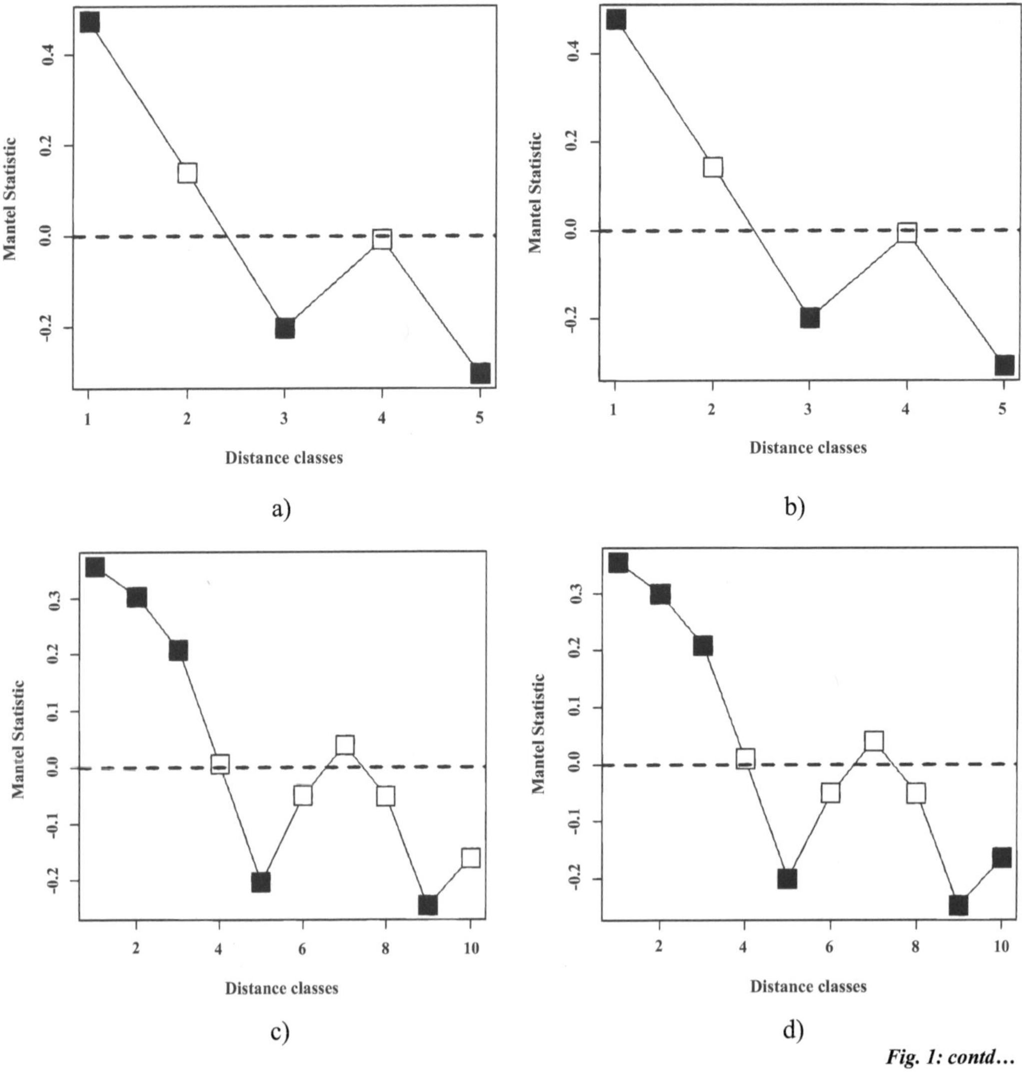

Fig. 1: contd...

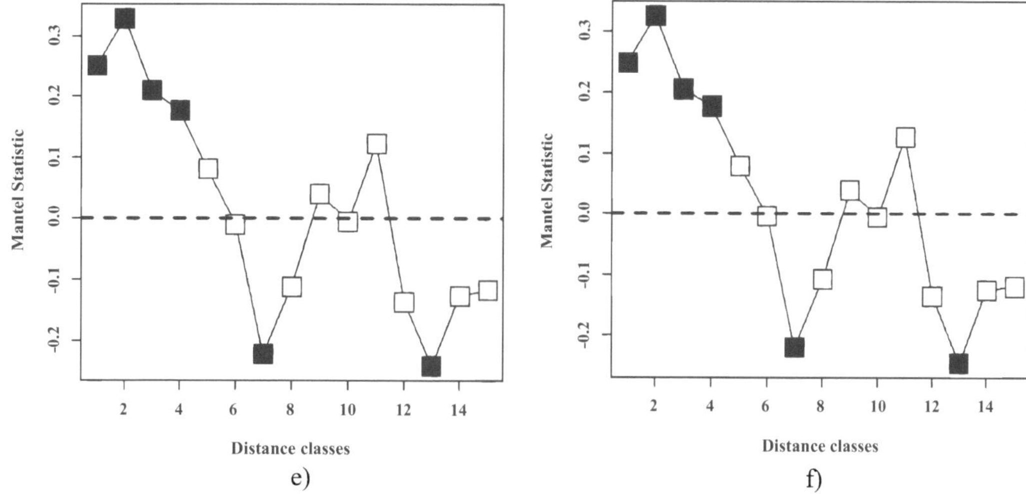

Figure 1: Mantel correlogram (a,b: distance number=5; c,d: distance number=10; e,f: distance number=15) for multiscale analysis of distributional concordance of birds and mammals in Hainan Island. The influence of environment was ignored across different distance classes in a,c,d, while controlled in b,d,e. Black solid squares denote significant permutation test ($P<0.01$).

Interestingly, we found that the influence of environment on distributional concordance of species tend to diversify the distributional difference of birds and mammals. This could be revealed by the partial Procrustes statistics in either a global test or multiscale correlograms. For the global test, the removal of environment will reduce m_{12} from 0.1166 to 0.0785. Similarly, for multiscale

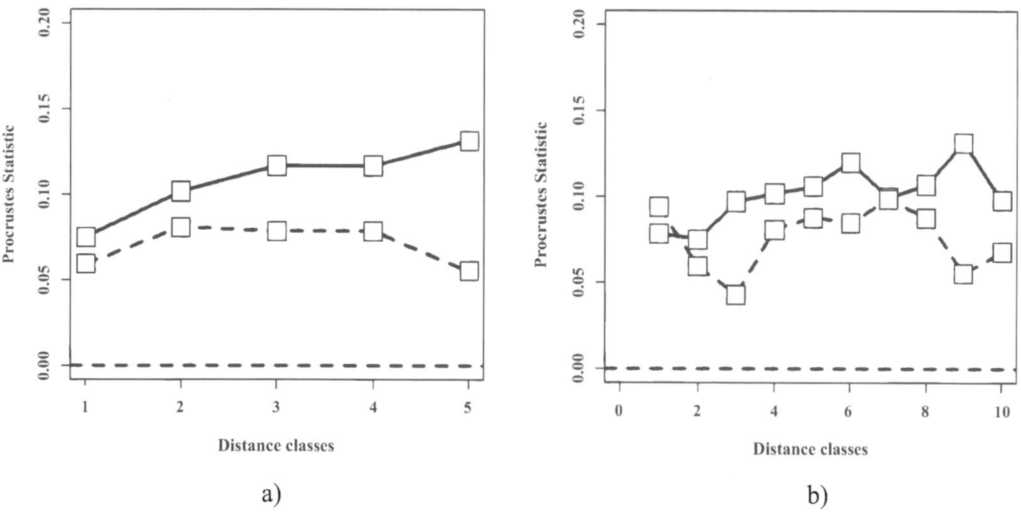

Fig. 2: contd…

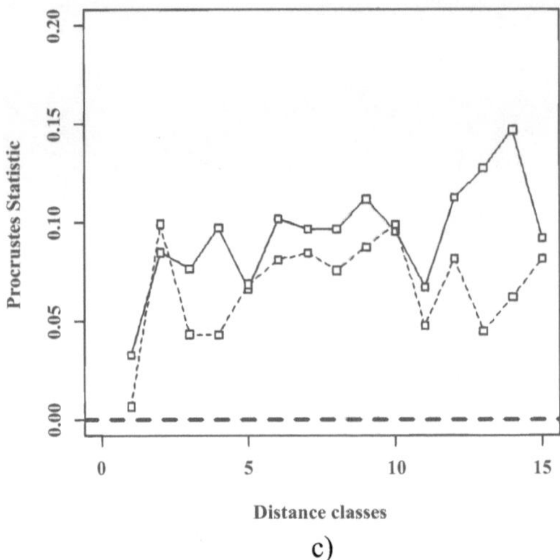

c)

Figure 2: Partial Procrustes correlogram (a: distance number=5; b: distance number=10; c: distance number=15) for multiscale analysis of distributional concordance of birds and mammals in Hainan Island. The influence of environment was either ignored (solid lines) or controlled (dashed lines) across different distance classes. Black solid squares denote significant permutation test (P<0.01).

correlograms, the partial cases after the controlling of environment usually have lower m_{12} statistics then those in the non-partial cases without controlling environment (Fig. **2**). In contrast, Mantel correlograms could not detect the sorting effects of environment on species' distribution. Mantel statistics didn't change in either environment-ignored or environment-controlled situations.

DISCUSSION

It has been revealed that the variance structure of transformed distance matrix (which is used in Mantel test) is different from the original species-site matrix, resulting into unpredictable results [4]. In contrast, the Procrustes test only relied on the original species-site matrix, avoiding the distance transformation and keeping the original variance property of the data. As such, it might be more appropriate of applying Procrustes test to reveal the distributional concordance pattern.

How could the Procrustes test maintain the original variance-covariance structure of the data? The simple reason is that the total variance for the whole matrix is given by $COV = (X - \overline{X})^T (X - \overline{X})$ and the variance for the sub matrix derived from the distance class d is $COV(d) = (X(d) - \overline{X}(d))^T (X(d) - \overline{X}(d))$. Because we know that the sub-matrix $X(d)$ is part of the whole matrix X, the variance and covariance inside $X(d)$ is a part of that for the whole matrix. So why does it need to discuss variance of the data for the spatial correlogram analysis? As seen in equations (4) and (5), the covariance structure of the data could influence the resultant Procrustes and Mantel statistics. Different to Mantel test relied on the distance matrices [4], Procrustes test directly manipulated the original species-site matrix without distance transformation, thus the resultant statistic could reflect the untransformed similarity of species-site matrices.

Partial correlograms, developed for Mantel and Procrustes statistics, could offer us an overall understanding of the multiscale patterns of matrix similarity by removing the effects of other factors. In our study, the distribution of amphibian and reptilian species in China was found to be determined by environmental conditions since the controlling of environmental influence could lead to insignificant results across different distance classes.

However, the utilization of partial analysis should be precautionary because the partial correlation analysis for Mantel statistic and the partial regression analysis for Procrustes statistic are two fundamentally different procedures. The results therefore could be totally misleading and contradictory. The reason is because the partial correlation depends on the manipulation of resultant covariance matrix for the three matrices X, Y and Z (the control matrix), while the partial regression analysis utilized the control matrix to regress against the response matrices X and Y. After that, the residual response matrices X_{res} and Y_{res} were subjected to further analysis. As seen, these two different procedures of course would generate different results on the partial analysis.

Furthermore, although Procrustes method could utilize original species-site matrix without data transformation, its application is still not fully desirable. As showed in Fig. **2**, Procrustes correlograms always have resultant Procrustes statistics closely related to the global value with same significant levels. Thus, multi-scale patterns of Procrustes statistics tended to be constant across distance classes. As such, the influence of spatial autocorrelation on structuring species distribution through Procrustes statistics was obscure and no large variations in Procrustes statistics could be found. From this perspective, Mantel statistic outperformed the Procrustes statistic, as it could detect the fluctuation of correlation of two distance matrices across different spatial distance classes.

This case study showed that the distribution of birds and mammals in Hainan Island was not structured by environment in principle, as evidenced by partial Mantel correlogram. However, environment did play a role on differentiating the distributional divergence of bird and mammal groups, as revealed by partial Procrustes correlogram. This contradictory result indicated the fundamental difference of Mantel and Procrustes tests as discussed above in detail. As such, both methods hold some advantages on revealing community structure of species distribution. Environment only played a weak role on the original species-site matrix in Hainan birds and mammals, but partial Mantel statistic did not find any evidences of correlation between environment and species presence-absence distribution. The derived distance matrices from environment and species-site matrix have converted the original community patterns to distance patterns for Mantel test, reducing the power to test the minor roles of environment on influencing species distribution.

In conclusion, both Mantel and Procrustes methods could help reveal different aspects of the varying-scale patterns on the similarity of species distribution. Moreover, their partial versions could further allow us to control the influence of other factors, observe the independent multi-scale relationship of species distribution, and test the role of the third factors. Procrustes method is effective because it doesn't require distance transformation and becomes sensitive to the influence of the third factors (*i.e.*, environment). However, it is insensitive to different spatial distances and the spatial autocorrelation patterns could not be analyzed. In contrast, Mantel method is effective on testing the influence of spatial autocorrelation and multiscale patterns of distance matrices could be revealed greatly.

REFERENCES

[1] J. Lamoreux, J. Morrison, T. Ricketts, D. Olson, E. Dinerstein, M. McKnight, *et al.*, "Global tests of biodiversity concordance and the importance of endemism", *Nature*. 440, 212-214, 2006.

[2] P. Legendre, L. Legendre, Numerical ecology, Elsevier Science BV, Amsterdam, 1998.

[3] P. Peres-Neto, D. Jackson, "How well do multivariate data sets match? The advantages of a Procrustean superimposition approach over the Mantel test", *Oecologia*. 129, 169-178, 2001.

[4] P. Legendre, M. Fortin, "Comparison of the Mantel test and alternative approaches for detecting complex multivarite relationships in the spatial analysis of genetic data", *Mol. Ecol. Resour.* 10, 831-844, 2010.

[5] D. Borcard, P. Legendre, "Is the Mantel correlogram powerful enough to be useful in ecological analysis? A simulation study", *Ecology*. 93, 1473-1481, 2012.

[6] N. Mantel, "The detection of disease clustering and a generalized regression approach", *Cancer Res.* 27, 209-220, 1967.

[7] D. Borcard, P. Legendre, P. Drapeau, "Partialling out the Spatial Component of Ecological Variation", *Ecology*. 73, 1045, 1992.

[8] H. Tuomisto, K. Ruokolainen, "Analyzing or explaining beta diversity? Reply", *Ecology*. 89, 3244-3256, 2008.

[9] P. Legendre, D. Borcard, P. Peres-Neto, "Analyzing beta diversity: Partitioning the spatial variation of community composition data", *Ecol. Monogr.* 75, 435-450, 2005.

[10] C. Wang, Z.A. Szpiech, J.H. Degnan, M. Jakobsson, T.J. Pemberton, J.A. Hardy, *et al.*, "Comparing Spatial Maps of Human Population-Genetic Variation Using Procrustes Analysis", *Stat. Appl. Genet. Mol. Biol.* 9, 13. doi:10.2202/1544-6115.1493, 2010.

[11] C. Wang, S. Zollner, N. Rosenberg, "A quantitative comparison of the similarity between genes and geography in worldwide human populations", *PloS Genet.* 8, e1002886, 2012.

[12] S. Matesanz, T. Gimeno, M. de la Cruz, A. Escudero, F. Valladares, "Competition may explain the fine-scale spatial patterns and genetic structure of two co-occurring plant congeners", *J. Ecol.* 99, 838-848, 2011.

[13] Perez-Ortega, R. Ortiz-Alvarez, T. Green, A. de los Rios, "Lichen myco- and photobiont diversity and their relationships at the edge of life (McMurdo Dry Valleys, Antarctica)", *FEMS Microbiol. Ecol.* 82, 429-448, 2012.

[14] J. Gower, "Generalized Procrustes analysis", *Psychometrika.* 40, 33-51, 1975.

[15] J. Gower, "A general coefficient of similarity and some of its properties", *Biometrics.* 27, 857-871, 1967.

[16] D. Jackson, "PROTEST: a Procrustean randomization test of community environment concordance", *Eoscience.* 2, 297-303, 1993.

[17] R Development Core Team, R: A Language and Environment for Statistical Computing, Vienna, Austria. ISBN 3-900051-07-0, URL http://www.R-project.org., (2013).

[18] J. Oksanen, G. Blanchet, R. Kindt, *et al.*, vegan: Community Ecology Package. R package version 2.0-4, (2012).

[19] R. Fisher, "On the "probable error" of a coefficient of correlation deduced from a small sample", *Metron.* 1, 1-32, 1921.

[20] W. Hardle, L. Simar, "Applied multivariate statistical analysis", Springer Berlin Heidelberg, 2012.

[21] H. Hotelling, "New light on the correlation coefficient and its transformation", *J. R. Stat. Soc. Ser. B.* 15, 193-232, 1953.

[22] Y. Chen, "Avian biogeography and conservation on Hainan Island, China", *Zoolog. Sci.* 25, 59-67, 2008.

[23] Y. Chen, "Distribution patterns and faunal characteristic of mammals on Hainan Island of China", *Folia Zool.* 58, 372-384, 2009.

Chapter 6: Ecological Ordination Methods-Principal Component Analysis, Principal Coordinate Analysis, Redundancy Analysis and Canonical Correspondence Analysis

CHAPTER 6

Ecological Ordination Methods-Principal Component Analysis, Principal Coordinate Analysis, Redundancy Analysis and Canonical Correspondence Analysis

Abstract: Ordination analyses played a central role in community ecology. Ordination of the objects, usually species or sites in ecological studies, is simply to apply multivariate statistics to reduce the dimensions of the data so as to quantify the major ecological trends hidden in the ecological data in terms of some ecological matrices, like species-site matrix and site-environment matrix. In this chapter, I presented some most widely used ordination methods and the associated statistical backgrounds. Interested readers into the details and other ordination methods that are not listed here should refer to the classical numerical ecology book written by Prof. Legendre [1].

Keywords: Canonical correspondence analysis, computational ecology, correlation analysis, dimension reduction, ecological gradients, environmental correlates, numerical ecology, principal component analysis, species communities, species/site ordering, statistical ecology.

INTRODUCTION

Ecological data typically is presented in matrix form, in which each column or row represented one dimension. A matrix could has many rows and columns. Thus, ecological data is usually in high dimensions.

High-dimension data are not directly visible or easily understood. Thus, it is very necessary to conduct some sorts of analyses to reduce the high-dimensional data into low-dimensional data (typically in two or three dimensions). The mapping of these low-dimensional data (or scores) in scatter plots thus presented the principal ecological gradients and reflected the projections of associations of the objects (*e.g.*, species or sites) in the high-dimensional space in low-dimensional space.

PRINCIPAL COMPONENT ANALYSIS (PCA)

PCA reduces attribute space from a larger number of variables to a smaller number of factors and as such is a "non-dependent" procedure. It should be noted that there is no guarantee that the reduced dimensions after PCA analysis are interpretable.

Youhua Chen

PCA analysis has been broadly applied in biodiversity patterns, especially when ones want to extract major environmental and geometric gradients of the spatial diversity patterns. PCA helps to remove the possible influence of highly correlated variables (that is, multi-collinearity problem).

Here is the mathematical background of PCA:

For a species-site $m \times n$ matrix *X*, it is required to compute variance-covariance (or distance) site matrix as follows:

$$C{=}Cov(X) = E[(X - E(X))(X - E(X))^T] \quad (1)$$

$$D_{ij} = \| x_i - x_j \|_2^2 \quad (2)$$

where $E(X)$ is the column means of the matrix *X*, x_i denotes the *i*-th column in the matrix *X*. $\|\bullet\|_2$ denotes the 2-norm on the hyper-dimensional Real space R^m.

Please be noted *Cov(X)* denotes the covariance matrix, for the covariance value between two vectors {*x*} and {*y*} (length of both is *n*), its computational formula is given by,

$$Cov(x, y) = \frac{\sum_{i=1}^{n} (x_i - \overline{x})(y_i - \overline{y})}{n-1} \quad (3)$$

where $\overline{x}$ and $\overline{y}$ denote the mean of the corresponding vectors.

PCA is to get corresponding eigenvalues and eigenvectors from the variance-covariance matrix *C* or distance matrix *D*, the eigenvalue decomposition on *D* is demonstrated as follows (PCA on the variance-covariance matrix *C* follows the same procedure),

The computation of eigenvalues is to solve the following equation,

$$|D\text{-}\lambda \mathrm{I}|=0 \quad (4)$$

where *I* is the identity matrix, $|\bullet|$ is the determinant operator, λ is the eigenvalue that is required to estimate.

The corresponding eigenvector u_k for a specific eigenvalue λ_k should satisfy,

$$Du_k = \lambda_k u_k \tag{5}$$

Thus, if λ_k is known, we can compute the corresponding eigenvector. Please be noted that there is no unique solution for u_k in the above equation. It is a direction vector only and can be scaled to any magnitude.

PRINCIPAL COORDINATE ANALYSIS (PCOA)

PCoA is also widely used in ecological studies to reduce the high-dimension data into low-dimension gradients. PCoA is very simple to compute, the only care is to centralize the distance matrix.

The computation of PCoA is illustrated as follows [1, 2]:

(1) The ecological matrix (*e.g.*, species-site matrix) should be converted into a distance matrix $D = [D_{ij}]$.

(2) The distance matrix is then transformed into a new matrix A the element which is calculated as,

$$a_{ij} = -\frac{1}{2} D_{ij}^2 \tag{6}$$

(3) The matrix A is then centralized to obtain a new matrix $\Delta_{ij} = [\delta_{ij}]$, in which δ_{ij} is computed as,

$$\delta_{ij} = a_{ij} - a_{\bullet j} - a_{i\bullet} + a_{\bullet\bullet} \tag{7}$$

where $a_{\bullet j}$ and $a_{i\bullet}$ are the column and row means for the column j and row i respectively; $a_{\bullet\bullet}$ is the overall mean of all elements in matrix A.

If we write in matrix form, the computation of the centralized matrix Δ is,

$$\Delta = (I - \frac{11^T}{n}) A (I - \frac{11^T}{n}) \tag{8}$$

where I is the identity matrix, T denotes the matrix transpose. 1 is a column vector with all elements inside being 1. n is the number of objects.

(4) The eigenvalues λ_k and eigenvectors u_k then are then computed. The computation of eigenvalues is to solve the following equation,

$$|\Delta\text{-}\lambda \mathrm{I}|=0 \tag{9}$$

where I is the identity matrix, $|\bullet|$ is the determinant operator, λ is the eigenvalue that is required to estimate.

The corresponding eigenvector u_k for a specific eigenvalue λ_k should satisfy,

$$\Delta u_k = \lambda_k u_k \tag{10}$$

Thus, if λ_k is known, we can compute the corresponding eigenvector in column form. Please be noted that there is no unique solution for u_k in the above equation. It is a direction vector only and can be scaled to any magnitude. For PCoA, the eigenvectors are scaled to lengths equal to the square roots of the respective eigenvalues as,

$$\sqrt{u_k^T u_k}=\sqrt{\lambda_k} \tag{11}$$

Please note that, due to the centralization of the matrix Δ, there is at least one zero eigenvalue.

(5) After scaling, if the eigenvectors are wrriten in columns, then the rows of the resulting table are the coordinates of the objects in PCO reduced space.

CANONICAL CORRESPONDENCE ANALYSIS (CCA)

CCA is also widely used in multivariate ordination analysis. The reason of utilizing CCA is that it can integrate the impact of environmental covariates on the ordination of the species/site in low-dimensional space. In contrast, either PCA or PCoA can only operate onto a single matrix (*e.g.*, species-site matrix) and make the corresponding ordination for the objects by assuming that the corresponding environment across different sites is homogeneous and constant. Such an implicit assumption clearly is not valid in real-world situations. Thus, it is very necessary to integrate the heterogeneous information conveyed by the environmental gradients where the sites are sampled. Thus, CCA is the ideal statistical technique to do such a work.

Assuming that matrices Y and X represented the response site-species abundance matrix ($n\times p$) and explanatory site-environment covariate matrix ($n\times q$). Then, the CCA analysis requires some transformation on the both response and explanatory matrices [3].

For matrix *Y*, the Chi-square transformation is done as follows,

$$Y'_{ij} = [y'_{ij}] = \frac{y_{ij} y_{\bullet\bullet} - y_{i\bullet} y_{\bullet j}}{y_{\bullet\bullet}\sqrt{y_{i\bullet} y_{\bullet j}}} \tag{12}$$

where $y_{\bullet j}$ and $y_{i\bullet}$ are the column and row means for the column *j* and row *i* respectively; $y_{\bullet\bullet}$ is the overall mean of all elements in matrix *Y*.

For matrix *X*, the transformation is done by weighting each row as follows,

$$X'_{ij} = [x'_{ij}] = x_{ij} / \sqrt{p_{i\bullet}} \tag{13}$$

where $p_{i\bullet} = y_{i\bullet} / y_{\bullet\bullet}$.

Then, the fitted values for response matrix $\hat{Y}'$ is given by,

$$\hat{Y}' = X'[X'^{T} X']X'^{T} Y' \tag{14}$$

where *T* denotes the matrix transpose. Y' and X' are the transformed forms for *Y* and *X* as just mentioned above.

Then covariance matrix for $\hat{Y}'$ is given by,

$$Q = \hat{Y}'^{T} \hat{Y}' \tag{15}$$

please note that the matrix size of *Q* is $p \times p$.

The standard eigenvalue decomposition as mentioned in PCA and PCoA analyses is then applied to the covariance matrix *Q* to obtain the corresponding eigenvalues *S* and eigenvectors *U*. Each column *i* of *U* has the corresponding eigenvalue S_i.

The corresponding eigenvector matrix *U* (each eigenvector in a column) should be scaled by the column weights of the response matrix *Y* to obtain the species scores *V* for scatter plotting as,

$$V_{ij} = [v_{ij}] = u_{ij} / \sqrt{p_{\bullet j}} \tag{16}$$

where u_{ij} is the element in eigenvector matrix *U*. $p_{\bullet j}$ is the column weights which is defined as, $p_{\bullet j} = y_{\bullet j} / y_{\bullet\bullet}$.

The corresponding site scores W is given by,

$$W_{ij} = [w_{ij}] = g_{ij} / \sqrt{p_{i\bullet}} / \sqrt{S_j} \tag{17}$$

where g_{ij} is the element in matrix G which is calculated by projecting the transformed Y matrix onto the reduced species eigenvector matrix U,

$$G = Y'U\,. \tag{18}$$

REDUNDANCY ANALYSIS (RDA)

RDA is very similar to CCA,. but it is much simpler. It can be called as multivariate regression analysis. Similar to CCA, assuming that matrices Y and X represented the response site-species abundance matrix ($n \times p$) and explanatory site-environment covariate matrix ($n \times q$). Then, the RDA is computed as the way to get the fitted values for the response variable matrix as,

$$\hat{Y} = X_1[X_1^T X_1]X_1^T Y \tag{19}$$

where X_1 is a modified version of X in which a column of 1 is added onto the matrix X (for computing intercepts of the multivariate regression).

Then, the covariance matrix Q based on the fitted values $\hat{Y}$ is calculated as,

$$Q = \frac{1}{n-1}\hat{Y}^T\hat{Y}\,. \tag{20}$$

Then, a PCA should be done on the matrix Q so as to obtain corresponding eigenvalues λ_k and eigenvectors u_k.

The eigenvectors should be scaled so as to obtain the species and site scores (the ordination of the species and sites in the reduced space).

IMPLICATIONS

The methods presented here have been widely applied in numerous empirical studies. The applications of these methods are the only ways to make your research become publishable. Because there are a number of open and commercial software that can implement these multivariate statistical methods, it is strongly encouraged that the readers of the book should try to practice these methods frequently in their research.

REFERENCES

[1] P. Legendre, L. Legendre, Numerical ecology, Elsevier Science BV, Amsterdam, 1998.

[2] J. Gower, "A general coefficient of similarity and some of its properties", *Biometrics*. 27, 857-871, 1967.

[3] V. Makarenkov, P. Legendre, "Nonlinear redundancy analysis and canonical correspondence analysis based on polynomial regression", *Ecology*. 83, 1146-1161, 2002.

Chapter 7: Variation Partitioning Techniques

CHAPTER 7

Variation Partitioning Techniques

Abstract: Variation partitioning is another important statistical method in numerical ecology studies. The purpose of variation partitioning is to decompose the variation presented by the species compositions across the studied sites into different categories. These categories of variation could be attributed to spatial covariates, environmental covariates, climatic covariates or others. In community ecology, spatial component of variation in the resultant ecological communities widely existed in ecological data sets. This is because spatial autocorrelation was a very natural phenomenon describing the similarity of neighboring objects (*e.g.*, species richness, species community structure or others). However, environmental gradients are also very dominant and widely observed in ecological community structure. Thus, what are the relative importance of environmental gradients and spatial autocorrelation on structuring ecological communities? Variation partitioning helps address this question.

Keywords: ANOVA, environment-trait relationship, F-statistics, Moran's I index, multivariate statistics, ordination analyses, spatial autocorrelation, spatial statistics, species distribution and diversity patterns, temporal autocorrelation, variation decomposition.

INTRODUCTION

Variation partitioning is a statistical method for helping elucidate the influence of each group of explanatory variables in multivariate statistics [1]. Variation partitioning has been broadly applied in ecological and evolutionary studies [2]. For simplicity, the mathematical formulation for variation partitioning technique is as follows [1, 3, 4].

In community ecology, variation partitioning was widely utilized to evaluate the relative importance of different ecological mechanisms on explaining the observed species compositional variation. Thus, it is recommended that the readers of the book should learn some basic backgrounds for this technique. In this chapter, I presented one statistical version of the variation partitioning method and the partitioning of beta diversity with null models so as to allow readers to know how the partitioning of beta diversity could be fulfilled using variation partitioning method.

VARIATION PARTITIONING ON SPECIES COMPOSITION MATRIX

Supposing that there are two groups of explanatory variables in two matrices X_1 and X_2, the total variation S in the response variable matrix Y with n rows is written as,

Youhua Chen

$$S = \frac{1}{n-1} Trace((Y - \overline{Y})^T (Y - \overline{Y}))$$

where a hyphen above the variable(s) denotes the mean(s).

Then the proportion of variation R_1 only explained by the group of explanatory variables X_1 is obtained as,

$$Y_1 = X_2 [X_2{}^T X_2]^{-1} X_2{}^T Y, \quad (1)$$

$$Y_1^{res} = Y - Y_1, \quad (2)$$

$$X_1{}^{res} = X_1 - X_2 [X_2{}^T X_2]^{-1} X_2{}^T X_1, \quad (3)$$

$$\hat{Y}_1 = X_1^{res} [X_1^{resT} X_1^{res}]^{-1} X_1^{resT} Y_1^{res}, \quad (4)$$

$$R_1 = Trace(\frac{1}{n-1} (\hat{Y}_1 - \overline{Y}_1)^T (\hat{Y}_1 - \overline{Y}_1)) \Big/ S \,. \quad (5)$$

the percentage of total variation R_2 attributed to the explanatory variable group X_2 is calculated following the same procedure as above.

Finally, the percentage of total variation explained by the interaction of the two variable groups X_1 and X_2 requires the determination of the percentage of variation (R_{12}) explained by all the variables *X*, the matrix of which combines matrices X_1 and X_2 together:

$$Y_{12} = X[X^T X]^{-1} X^T Y, \quad (6)$$

$$R_{12} = Trace(\frac{1}{n-1} (Y_{12} - \overline{Y}_{12})^T (Y_{12} - \overline{Y}_{12})) / S \,. \quad (7)$$

Thus, the proportion of variation that can not explained by any current explanatory variables is determined by,

$$R_0 = 1 - R_{12} \,. \quad (8)$$

Then, the percentage of total variation explained by the interaction of the two variable groups is given by,

$$R_{1\cap 2} = R_{12} - R_1 - R_2 \tag{9}$$

In a summary, R_1, R_2, $R_{1\cap 2}$ and R_0 are the focused explained variation for the present study. Fig. **1** showed the diagram for the relationships of R_1, R_2, $R_{1\cap 2}$ and R_0.

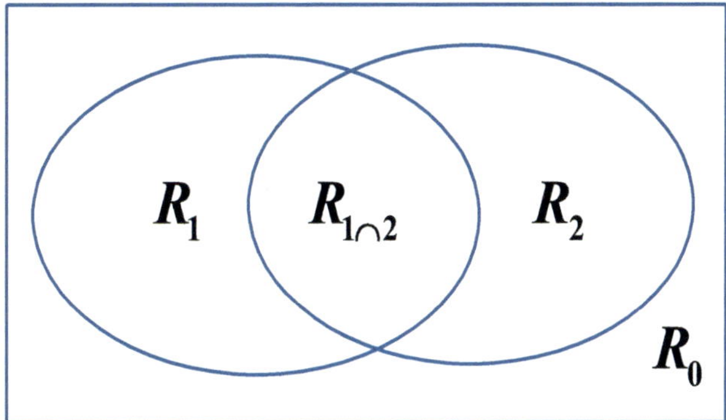

Figure 1: A diagram showing the relationships between the variation components R_1, R_2, $R_{1\cap 2}$ and R_0.

Such a variation partitioning procedure could be extended to the situation where there were three or more groups of variables.

VARIATION PARTITIONING ON BETA DIVERSITY WITH NULL MODEL

Measurement of Beta Diversity

Following some previous studies [5, 6], I quantified and modeled beta-diversity as the way like this: the observed beta-diversity of a community defined as follows:

$$B_{\text{obs}} = Var(Y)/(n-1) \tag{10}$$

where Y is the Hellinger transformation of the original abundance-site matrix X for a focused plot:

$$Y_{ij} = \sqrt{X_{ij} / \sum_j X_{ij}} \tag{11},$$

and $Var(Y) = Trace(YY^T)$. T denotes the transpose of a matrix, *Trace()* is the sum of the diagonal elements of a matrix.

The expected beta-diversity of the studies moss plot are generated by a null model [6]. During the randomization, the total individual per site and the relative abundance of species in the species pool were kept unchanged. The beta-diversity was calculated for each simulated random matrix using equation (1) after Hellinger transformation as above. The mean of the simulated beta-diversity is taken as expected beta-diversity B_{exp}. Finally, the deviation of beta-diversity B_{dev} is quantified as the difference of observed and expected beta-diversity, divided by the standard deviation of expected beta-diversity [6]. The deviation of beta-diversity is treated as the focused beta-diversity in which the random sampling effect has been removed. Fig. **2** showed the relationships between B_{obs}, B_{exp} and B_{dev}.

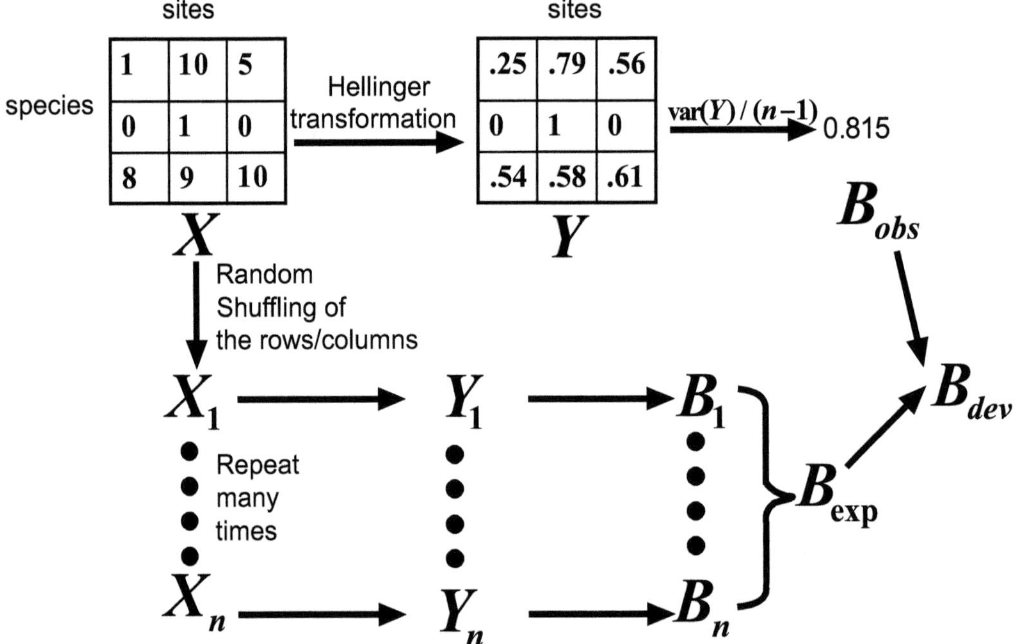

Figure 2: The computation of different beta diversity metrics.

The partitioning of variation in the species composition data matrix X is implemented using partial canonical correspondence analysis (pCCA) [7]. The two categories of explanatory variables, micro-environmental variable matrix E and macro-climatic variable matrix C, are used as covariance matrices for the purpose to determine the relative contribution of long-term and short-term ecological processes on structuring beta diversity patterns of microarthropod communities.

Proportions of beta diversity that are explained by micro-environmental variables and macro-climatic variables, after the controlling of random sampling effect, thus are quantified as follows [6],

$$B_E = B_{dev} \times R^2_{adj}(E) \tag{12}$$

$$B_C = B_{dev} \times R^2_{adj}(C) \tag{13}$$

here, $R^2_{adj}(E)$ and $R^2_{adj}(C)$ denote the proportions of variation of species composition explained by micro-environmental and macro-climatic variables solely respectively.

Finally, the proportion of beta diversity that is explained by the interaction of micro-environmental and macro-climatic variables, after the controlling of random sampling effect, is defined as,

$$B_{E \times C} = B_{dev} \times R^2_{adj}(E \times C) \tag{14}$$

Similarly, $R^2_{adj}(E \times C)$ denotes the proportion of variation of species composition jointly explained by micro-environmental and macro-climatic variables. Fig. **3** showed the relationship between $R^2_{adj}(E)$, $R^2_{adj}(C)$ and $R^2_{adj}(E \times C)$.

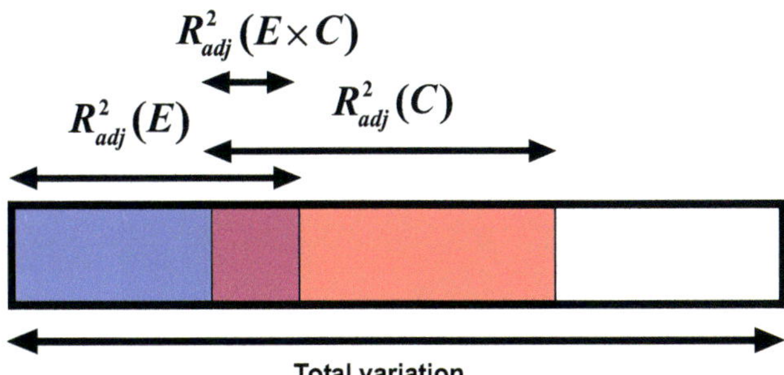

Figure 3: Variation partitioning and the relationship of different components of variation attributed to different groups of covariates.

REMARKS

Variation partitioning has been widely applied in community ecology to disentangle the relative roles of niche and neutrality processes on structuring community composition and beta diversity patterns of species assemblage. There are other ways to conduct variation partitioning, but the basic idea is similar.

REFERENCES

[1] P. Peres-Neto, P. Legendre, S. Dray, D. Borcard, "Variation partitioning of species data matrices: estimation and comparison of fractions", *Ecology*. 87, 2614-2625, 2006.

[2] P. Legendre, L. Legendre, Numerical ecology, Elsevier Science BV, Amsterdam, 1998.

[3] P. Legendre, D. Borcard, P. Peres-Neto, "Analyzing beta diversity: Partitioning the spatial variation of community composition data", *Ecol. Monogr*. 75, 435-450, 2005.

[4] D. Borcard, P. Legendre, P. Drapeau, "Partialling out the Spatial Component of Ecological Variation", *Ecology*. 73, 1045, 1992.

[5] N. Kraft, L. Comita, J. Chase, N. Sanders, N. Swenson, T. Crist, *et al.*, "Disentangling the drivers of beta diversity along latitudinal and elevational gradients", Science. 333, 1755-1758, 2011.

[6] M. De Caceres, P. Legendre, R. Valencia, M. Cao, L. Chang, "The variation of tree beta diversity across a global network of forest plots", *Glob. Ecol. Biogeogr.* 21, 1191-1202, 2012.

[7] C.J.F. ter Braak, "Canonical correspondence analysis : a new eigenvector technique for multivariate direct gradient analysis", *Ecology*. 67, 1167-1179, 1986.

CHAPTER 8

Species-Area, Commonness-Area, Rarity-Area and Endemic Species-Area Relationships

Abstract: Modeling species-area and endemic species-area relationships have been broadly focused. However, little attention has been paid to rare species-area relationship and common species-area relationship. Also, there were few studies revealing the shape patterns among the curves. In the present study, we quantified the relationships among species-area (SAR), commonness-area (CAR), rarity-area (RAR) and endemism-area (EAR) relationships by investigating different probability functions of species distribution (including negative binomial distribution, finite negative binomial distribution, Poisson distribution, geometric distribution and He-Legendre extension of negative binomial distribution). Results showed that RAR could be used as a robust indicator of SAR, because the curve patterns are identical between RAR and SAR in most cases. CAR could easily become saturated for all probability models in that common species are readily detectable. Finally, EAR curves have convex patterns compared to other relationships. The influences of species aggregation and abundance on SAR and EAR typically showed opposite patterns: increasing aggregation and reducing abundance would lower SAR curves, while uplift EAR curves. Interestingly, varying aggregation and abundance of species could not change the flat patterns of CAR and predict the curve shapes of RAR among different probability models.

Keywords: Aggregation, commonness, ecological universal rules, endemism, flagship species, geometric constraint, indicator species, power law, rarity, sampling bias, species distribution, species-area relationship.

INTRODUCTION

Aggregation of species distribution is ubiquitous in natural world [1-3]. Predicting species occurrence using aggregated mathematical models should be superior compared to those without aggregation effects. In modeling aggregated distribution of species, two common probability distributions, Poisson distribution [4] and negative binomial distribution [1-3, 5, 6] were widely applied.

There were several advances on developing aggregation models of species. For example, Zillio and He [1] proposed a finite negative binomial distribution model for better assessing the aggregation coefficient k for rare species. These rare species typically were said to be hard to converge when fitting maximum likelihood for negative binomial distribution [1].

One important application of species abundance distributional models is to model the species-area relationship [1, 7-11], and recently, to model the endemic

Youhua Chen

species-area relationship [8, 12]. Endemic species-area relationship has now been widely recognized for its importance for predicting extinction [8, 12], assessing the impacts of global changes [13] and so on. However, up to date, we do not know the impacts of different abundance models on changing the curve patterns of species-area and endemic species-area relationships. The interesting aspects for this question are that when we could understand the curve patterns driven by the different factors, we might be able to predict and estimate the biodiversity loss in a better way [8]. Although the comparison between endemic species-area and species-area relationships has been revealed clearly in the work of Green and Ostling [12], the patterns involved in common species-area and rare species-area relationships have never been studied.

In the present study, I made an attempt to compare different species-area relationships (species-area, endemic species-area, common species-area and rare species-area) generated by different probability models and made a simple comparison of the curve patterns and assessment on how the aggregation coefficient and species abundance would influence these curves.

MATERIALS AND METHODS

Species-Area Relationship (SAR)

Negative binomial distribution model and finite version for describing aggregation patterns of species are given by the following two equations respectively,

$$P_{NBD}(n) = \binom{n+k-1}{n}\left(\frac{Na}{Na+k}\right)^n\left(\frac{k}{Na+k}\right)^k \tag{1}$$

$$P_{FNBD}(n) = \frac{\binom{n+k-1}{n}\binom{N-n+k/a-k-1}{N-n}}{\binom{N+k/a-1}{N}} \tag{2}$$

Poisson model, however, does not count on the aggregation coefficient *k*, as,

$$P_{Pois}(n) = \frac{(Na)^n e^{-Na}}{n!} \tag{3}$$

for all models, n=0,1,2,...

Here, a is the proportion of areas being sampled from the total area A.

The density probability for different models with individuals larger than zero should be,

$$P_{NBD}(n>0\,|\,a)=1-\left(\frac{k}{Na+k}\right)^{k} \tag{4}$$

$$P_{FNBD}(n>0\,|\,a)=1-\frac{\Gamma(N+k/a-k)\Gamma(k/a)}{\Gamma(N+k/a)\Gamma(k/a-k)} \tag{5}$$

$$P_{Pois}(n>0\,|\,a)=1-e^{-Na} \tag{6}$$

it is noted that the density probability for Poisson model is also the simplest model of occupancy (see [14]). The NBD model $P_{NBD}(n>0\,|\,a)$ is also present in [15].

Further, we considered an additional probability function from [10, 12] as follows,

$$P_{HL}(n>0\,|\,a)=1-(1-a)(1+Na/k)^{-k} \tag{7}$$

This equation has been extensively tested in the work of [12], it is noted that the HL probability function $P_{HL}(n>0\,|\,a)$ was quite similar to NBD function $P_{NBD}(n>0\,|\,a)$, but with an extra term (1-a). Thus, I called this function as He-Legendre extension of NBD probability function (HL function).

So, the species-area relationship generated from the density probability given the sampling area proportional to the total area with parameter a, then we could have [1],

$$\langle s\rangle_a=\sum_{i=1}^{S}P(n>0\,|\,a) \tag{8}$$

here S is the total species number in the total area.

Common Species-Area Relationship (CAR)

For exploring common species-area relationship, we should firstly consider the definition of common species. Both distributional range and population density could be used to define species commonness and rarity. For modeling species-area relationship, it seemed hard to obtain distribution range of species. However, species abundance is one of the prerequisites to estimate distributional probability functions. Thus, we should consider the species in the community with abundances larger than a given threshold as the definition of common species. In contrast, those species with abundances less than a given threshold could be used to define rare species.

Here we choose the common species which have population sizes larger than 70% of the maximum population size (from a certain species) in the community. The selection of the threshold values could be a bit arbitrary and wouldn't influence the results.

The calculation of theoretical common species-area relationships is identical to those present in species-area relationship, but we only sum the probabilities of common species. Thus,

$$\langle s \rangle_a = \sum_{i=1}^{S_C} P(n > 0 \mid a) \tag{9}$$

where S_C is the number of common species in the community.

Rare Species-Area Relationship (RAR)

As discussed above, the definition of rare species should be those with population sizes less than a given threshold. Here, we could assign the threshold values 30% to select those species with population sizes less than 30% of the maximum population size (from a certain species) in the community. The selection of the threshold values could be a bit arbitrary and wouldn't influence the results.

The calculation of rare species-area relationships is identical to those present in species-area relationship, but we only sum the probabilities of rare species. Thus,

$$\langle s \rangle_a = \sum_{i=1}^{S_R} P(n > 0 \mid a) \tag{10}$$

where S_R is the number of rare species in the community.

Endemic Species-Area Relationship (EAR)

The probability that a species would be endemic to the area is the same as the probability that it would be absent from the other area which has the percentage of *(1-a)*, so the probability should be [12],

$$P^{E}_{NBD}(n>0\,|\,a)=1-P_{NBD}(n>0\,|\,1-a)=\left(\frac{k}{N(1-a)+k}\right)^{k} \tag{11}$$

$$P^{E}_{FNBD}(n>0\,|\,a)=1-P_{FNBD}(n>0\,|\,1-a)=\frac{\Gamma(N+k/(1-a)-k)\Gamma(k/(1-a))}{\Gamma(N+k/(1-a))\Gamma(k/(1-a)-k)} \tag{12}$$

$$P^{E}_{Pois}(n>0\,|\,a)=1-P_{Pois}(n>0\,|\,1-a)=e^{-N(1-a)} \tag{13}$$

$$P^{E}_{HL}(n>0\,|\,a)=1-P_{HL}(n>0\,|\,A(1-a))=a(1+N(1-a)/k)^{-k} \tag{14}$$

Thus, for endemic species, the density probability with individuals being large than zero in the sample site with proportion of a, should be,

$$\langle e\rangle_{a}=\sum_{i=1}^{S}P^{E}(n>0\,|\,a) \tag{15}$$

As the comparison, a geometric model of species-area and endemic species-area relationships without aggregation patterns of species distribution were also considered as follows [10, 12],

$$\langle s\rangle_{a}^{Geo}=S-\sum_{i=1}^{S}(1-a)^{N_i} \tag{16}$$

$$\langle e\rangle_{a}^{Geo}=\sum_{i=1}^{S}a^{N_i} \tag{17}$$

Thus, in summary, we have 3 aggregation probability functions (NBD, FNBD, HL) and two aggregation-free probability functions (Geo and Pois).

Numerical Simulations

For making simulations, we considered that there were 1000 species in the sample pool, and the population distribution of each species is assumed to follow the geometric distribution with mean abundance=100:

$$N_i \sim Geometric(0.01) \tag{18}$$

The aggregation parameter distribution of each species is assumed to have the Poisson probability function (add 1 is to exclude zeros):

$$K_i \sim Poisson(5) + 1 \quad (19)$$

Selection of different probability functions to model species abundance and aggregation patterns wouldn't influence the resultant curve patterns for different relationships.

After making a holistic comparison of SAR, CAR, RAR and EAR patterns, we chose NBD probability function to perform further analysis. Firstly, we decreased mean species aggregation while fixing abundance of species to see the influence of species aggregation on changing different curves. Secondly, we decreased mean species abundance while fixing aggregation of species to see the influence of species abundance on changing different curves;

RESULTS

Overview of SAR, CAR, RAR, and EAR Curves

The SAR, CAR, and RAR curves showed the concave patterns, in contrast, EAR showed linear curves, regardless to any specific probability distribution functions (Fig. **1**).

Interestingly, we found that the SAR curves for different probability functions are hard to distinguish when we consider a high aggregation parameter (mean=5) and species population size (mean=1000).

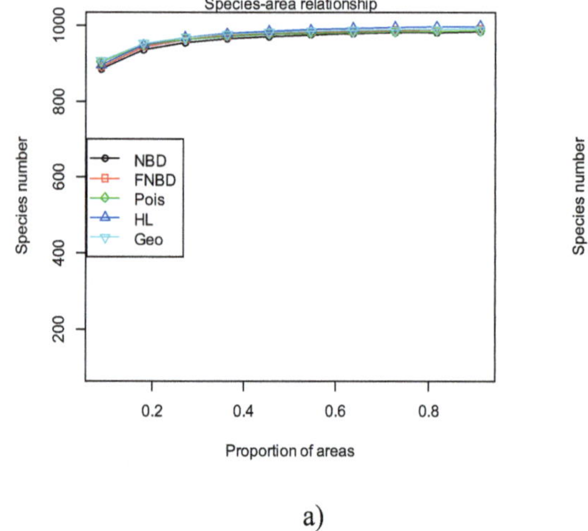

a)

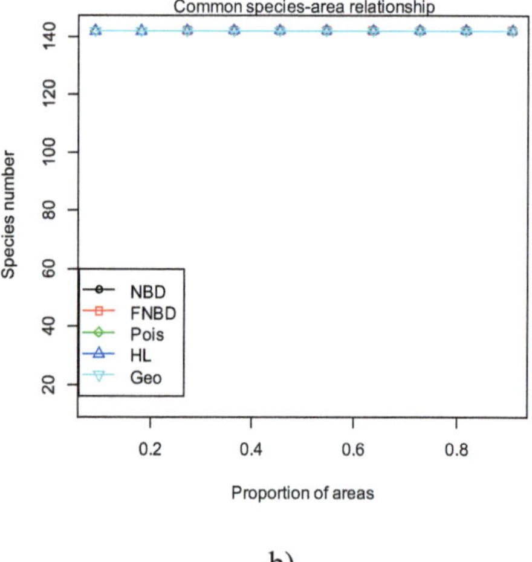

b)

Fig. 1: contd….

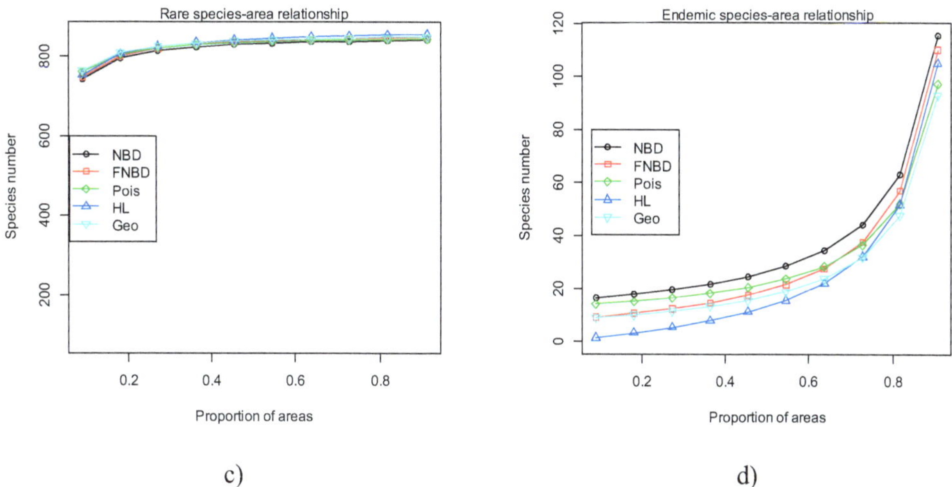

c) d)

Figure 1: Comparison of SAR, CAR, RAR, EAR curve patterns. Parameter settings: $N_i \sim Geometric(0.01)$ and $K_i \sim Poisson(5)+1$.

However, the SAR curves tended to separate from each other for the different probability density functions when we used geometric distribution to generate species population sizes. In specific, aggregation-free Poisson and Geometric distribution functions tended to have higher species number when sampling different proportions of areas (Fig. **1a**). In contrast, FNBD and NBD tend to have the lowest species numbers and both lines overlapped quite well. The SAR curve for HL probability function was located at intermediate positions among other curves.

With respect to EAR curves, that all probability functions have convex shapes. More importantly, we found that almost all the probability functions predicted flat EAR linear patterns except NBD model. NBD model showed a positive relationship between endemic species number and sampling area proportion. HL model seemed excluding endemic species, always having the lowest number of endemic species across different proportion of sampling areas and the values being around 0 when sampling area size is small (tested by different combinations of population sizes and aggregation parameter random variate generators).

Influence of Aggregation Patterns of Species on SAR, CAR, RAR, and EAR

Not surprisingly, increasing the magnitude of aggregation parameters of species would lead to higher species number in a given proportion of sampling area for

SAR curves (Fig. **2a**). This is because the reducing *k* values of species would lead to decreasing aggregation of species distribution, resulting in increasing detection probability of species. Correspondingly, EAR curves showed opposite patterns (Fig. **2g**), which showed that increasing k values of species would reduce species number in a given sampling size. This is simply the opposite logic for SAR. When aggregation of species reduced, the probability of species being common across different areas was enhanced.

Aggregation of species did not influence CAR patterns. All the curves from different probability functions tended to overlap together and were flat across different sampling effects (Fig. **2c**). In contrast, influence of aggregation on RAR was basically similar to those of SAR (Fig. **2e**). Thus, this result further verified our above statement that RAR curves largely affected SAR patterns.

Influence of Species Abundance on SAR, CAR, RAR and EAR

Increasing species abundance in the sampling area could also uplift the SAR curves (Fig. **2b**), because more species' individuals in the area, more likelihood of being commonly found. Reversely, increasing abundance would lower the EAR curves (Fig. **2h**), likely due to the fact that high abundance species tended to become range-wide and not endemic.

Interesting patterns were found on CAR and RAR. Increasing species abundance could not predict the shifting patterns of CAR and RAR curves. As showed in Fig. **2f**, for RAR curves, the communities with mean species abundance 25 and 20 would have accelerating increasing patterns when more areas were sampled, and could pass the RAR curve at some points (>0.5) with abundance 100. However, the RAR curves for abundance 50 and 33 were always laid under the CAR curve of abundance 100. For CAR curves, we found that the sampling effort could not increase the species number and the curves tended to be flat as all the common species could be detected even when sampling size is small, this result was consistent with the situation when aggregation of species was used as the independent factor (Fig. **2c**). At the meanwhile, increasing species abundance could not always increase species number for different sampling proportions of areas. CAR curve for abundance class 20 always had the highest number of species, while others (25, 33, 50) were laid under the one with the largest abundance 100 (Fig. **2d**).

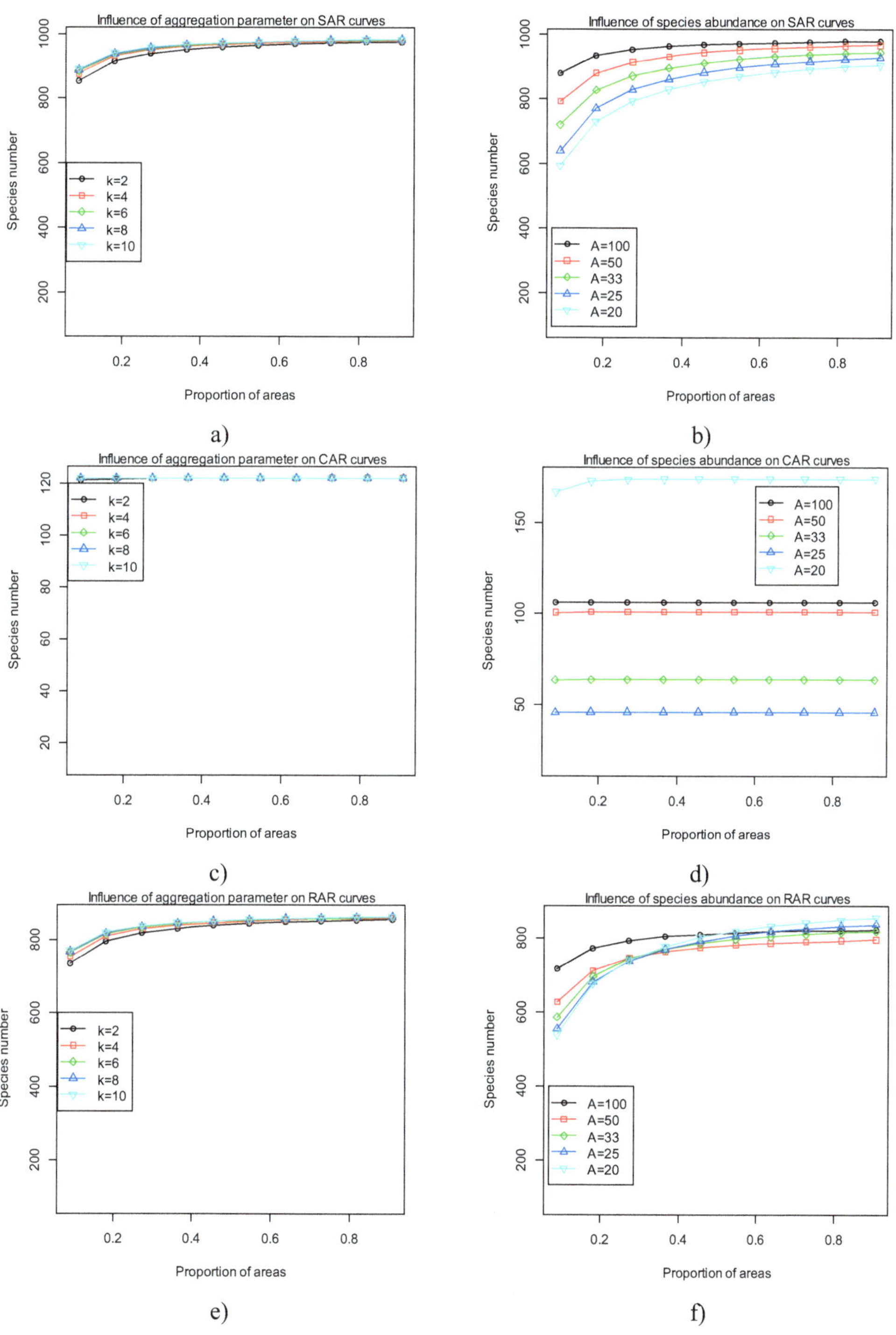
Influence of aggregation parameter on SAR curves
Species number
Proportion of areas
k=2
k=4
k=6
k=8
k=10
a)
Influence of species abundance on SAR curves
A=100
A=50
A=33
A=25
A=20
b)
Influence of aggregation parameter on CAR curves
c)
Influence of species abundance on CAR curves
d)
Influence of aggregation parameter on RAR curves
e)
Influence of species abundance on RAR curves
f)

Fig. 2: contd….

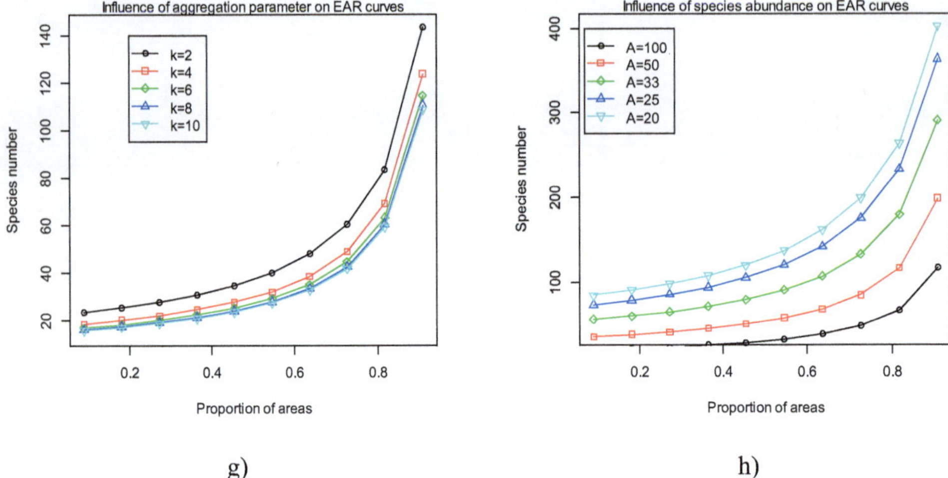

Figure 2: Influences of aggregation patterns and abundance of species on SAR, CAR, RAR, EAR curves. *k* indicated the mean aggregation parameter value for the species; *A* depicted the mean species abundance.

DSICUSSION

Interestingly, we found that for all the relationships, plots generated by negative binomial distribution were always located in the highest (for EAR) or lowest positions (for SAR, CAR, RAR) when sampling effort is sufficient. Thus, we argued that NBD model could be served as a basic model to compare either aggregated or not-aggregated species distribution patterns.

We found that RAR usually had identical curve patterns as those in SAR (Fig. **1a** and **1c**; Fig. **2a** and **2e**), although this result was not strongly supported when we modified species abundance (Fig. **2b** and **2f**). Whatever, we believed that SAR curve patterns could be largely driven by RAR. In contrast, we found that CAR curves could easily reach saturated patterns when sampling proportion of areas is further increasing (Figs. **1b**, **2c** and **2d**). Thus, CAR curves tended to be flat when sampling effort is enough to detect all common species (typically it is not hard since common species have very high probabilities of occurrence being detected). This is the main reason why RAR curve patterns could configure the basic shapes of SAR curve patterns.

Our results were different from pervious works, which showed that SAR curve patterns could have either convex or concave forms [16]. We could not observe convex patterns for SAR. A key reason accounting for this discrepancy might be due to the sampling difference of our work with the previous one: sampling of species in our study was spatially implicitly while the previous one used spatially explicitly sampling grid cells. The heterogeneity caused by spatial landscape should therefore create different scenarios of SAR scenarios. Another reason for the emergence of convex patterns in SAR might be due to the scaling and transformation of original species number or areas, which has been done in the work of (see Fig. **2** in [12]). In addition, all EAR curves showed convex patterns, consistent with the observation in [12].

Our results showed congruence aspects with the work in [16]. The roles of common and rare species were consistent among the two studies. SAR would be flattened when adding more common species in the community. However, the shape of SAR should be sharper when adding more rare species in the community (Fig. **2b**).

REFERENCES

[1] T. Zillio, F. He, "Modeling spatial aggregation of finite populations", *Ecology*. 91, 3698-3706, 2010.

[2] L. Taylor, I. Woiwod, J. Perry, "The density dependence of spatial behaviour and the rarity of randomness", *J. Anim. Ecol.* 47, 383-406, 1978.

[3] E. Pielou, Mathematical Ecology, Wiley, New York, 1977.

[4] P. Diggle, Statistical analysis of spatial point patterns, Academic Press, London, UK, 2003.

[5] J. Perry, L. Taylor, "Ades: new ecological families of species-specific frequency distributions that describe repreated spatial samples with an intrinsic power-law variance-mean property", *J. Anim. Ecol.* 54, 931-953, 1985.

[6] M. Boswell, G. Patil, Chance mechanisms generating the negative binomial distribution, in: G. Patil (Ed.), Random Counts Models Struct., Pennsylvania State University Press, University Park, Pennsylvania, USA, 1970: pp. 3-22.

[7] F. He, P. Legendre, "On species-area relations", *Am. Nat.* 148, 719-737, 1996.

[8] F. He, S. Hubbell, "Species-area relationships always overestimate extinction rates from habitat loss", *Nature*. 473, 368-371, 2011.

[9] B. Coleman, "Random placement and species-area relations", *Math. Biosci.* 54, 191-215, 1981.

[10] F. He, P. Legendre, "Species diversity patterns derived from species-area models", *Ecology*. 83, 1185-1198, 2002.

[11] O. Arrhenius, "Species and area", *J. Ecol.* 9, 95-99, 1921.

[12] J. Green, "A. Ostling, Endemics-area relationships: the influence of species dominance and spatial aggregation", *Ecology*. 84, 3090-3097, 2003.

[13] J. Malcolm, L. C, R. Neilson, L. Hansen, L. Hannah, "Global warming and extinctions of endemic species from biodiversity hotspots", *Conserv. Biol.* 20, 538-548, 2006.

[14] F. He, K. Gston, J. Wu, "On species occupancy-abundance models", *Ecoscience*. 9, 119-126, 2002.

[15] F. He, K. Gaston, "Estimating species abundance form occurrence", *Am. Nat.* 156, 553-559, 2000.

[16] E. Tjorve, W. Kunin, C. Polce, K. Tjorve, "Species-area relationship: separating the effects of species abundance and spatial distribution", *J. Ecol.* 96, 1141-1151, 2008.

Chapter 9: Species Abundance Distribution

CHAPTER 9

Species Abundance Distribution

Abstract: This chapter provides some models for quantifying species abundance distribution, a very basic but important pattern in ecological and biodiversity studies. The metrics for species abundance distribution covered only some common used ones in this book, like Zipf model, Broken stick model, Niche preemption model and Neutral model. An empirical data set, the species abundance distribution for Collembola, Mesostigmata and Oribatida found in Southwest of Canada was fitted by the above alternative models and the best model was selected through model comparison. Some simple methods for conducting model comparison have been referred in this chapter of the book.

Keywords: Akaike information criteria, biodiversity metrics, Canada, Collembola, ecological laws, Mesostigmata, mite ecology, model comparison and selection, niche versus neutrality, Oribatida, species abundance and distribution, species abundance distribution, statistical probability.

INTRODUCTION

Over the past ten years, there was a hot debate on the deterministic role of niche and neutrality processes on influencing biological communities [1-6]. Neutrality theory assumed no differentiation between species, thus each species is functionally and physiologically equivalent [3, 7]. Species under neutral theory have identical birth, death and mutation rates. Moreover, neutral theory emphasizes the importance of stochasticity, which has been usually overlooked in the niche theory [8].

The most striking support for neutral theory is the fitting of species abundance curve (SAD), which might yield the highest fit for the neutral model [9, 10]. However, other simple models, whether they were directly derived from niche theory or not, also have remarkably high fitting powers. Also, even when neutral theory could predict empirical SAD perfectly, the underpinning mechanisms driving species assemblages are still niche-based [8]. SAD could not fully reflect the mechanisms structuring species communities.

The comparison of different statistical models on their powers for fitting SAD has been well quantified in recent studies [11-14]. In most cases, simple statistical models are powerful enough to quantify SAD [8], and the neutral model didn't have a remarkably better fit.

Youhua Chen

INTRODUCING SOME SAD MODELS

Zipf Model

The existence of a new species in the community is influenced by the species arrived earlier [8], thus, the Zipf model [15] has the formula as,

$$N_i = Nqi^{-r} \tag{1}$$

where N_i denoted the predicted abundance for the *i*-th species in the SAD, N is the total individual number in the community, q is the predicted relative abundance of the species with highest abundance in the community, γ indicated the influence of priority effect.

Broken Stick Model (BSM)

BSM model [16] has the expected abundance for the *i*-th species based as below,

$$N_i = \alpha \sum_{k=i}^{S} 1/k \tag{2}$$

where α is the estimated scale parameter, S is the total species number in the community.

Niche Preemption Model (NPM)

NPM [17] assumed that the percentage of the total niche occupied by the first species is α, the second one occupied a percentage α of the reminder, being $\alpha(1-\alpha)$, and so on… As such, the expected abundance for the *i*-th species is,

$$N_i = N\alpha(1-\alpha)^{i-1} \tag{3}$$

Geometric Model (GEOM)

GEOM [18] is another form of niche preemption model, but the formula is different since it has two independent parameters,

$$N_i = \alpha\beta^{i-1} \tag{4}$$

Neutral Model (NM)

NM sampling formula is complex [19], for simplicity, it is not present here. Two important parameters, fundamental biodiversity index θ and migration rate *m*, are fitted using the program Tetame version 2.1 [20]. Expected abundance of species were estimated by taking the means of 1000 simulations of neutral communities using the estimated θ, m and total individual number J (for the whole dataset, J=13260) as the input in "urn.gp" program [19] under the PARI computational algebra system (http://pari.math.u-bordeaux.fr/).

MODEL COMPARISON AND EVALUATION

There are multiple statistical methods for testing the alternative SAD models. Here, we introduced three major methods: χ^2 test, Kolmogorov-Smirnov (KS) test and the Akaike Information Criterion (AIC) method is used as well to compare the models and identify the bets model by using log-likelihoods (log L) of the fitted models as the input. The calculation of AIC formula is given by,

$$AIC = -2\log L + 2k \tag{5}$$

where k is the parameter number in the fitted model.

A PRACTICAL EXAMPLE

Sampling Locations

32 moss field plots were surveyed across Southwest Canada based on the following standards of site selection during the summer time between the years of 2011 and 2012: (1) they should be contiguous with the mainland (islands excluded); (2) they should be flattened large rocky outcrops with > 4m^2 of moss carpets; (3) they should be accessed easily, being adjacent to highway roads. 353 morphospecies were identified and a total number of 13260 individuals were counted in the lab. Only the species belonging to the following orders would be considered in the study: Oribatida, Mesostigmata and Collembola [21]. The abundance of each species was calculated and utilized in the subsequent analyses.

SAD for Microarthropod Species as a Whole

As showed in Fig. **1** and Table **1**, all the models could fit the whole-microarthropod SAD quite well, but the difference between the expected and observed SAD still have a large significant discrepancy (indicated by χ^2 and K-S tests). Among the models, Zipf model has the lowest AIC value, indicating the most favored model for whole-microarthropod SAD pattern.

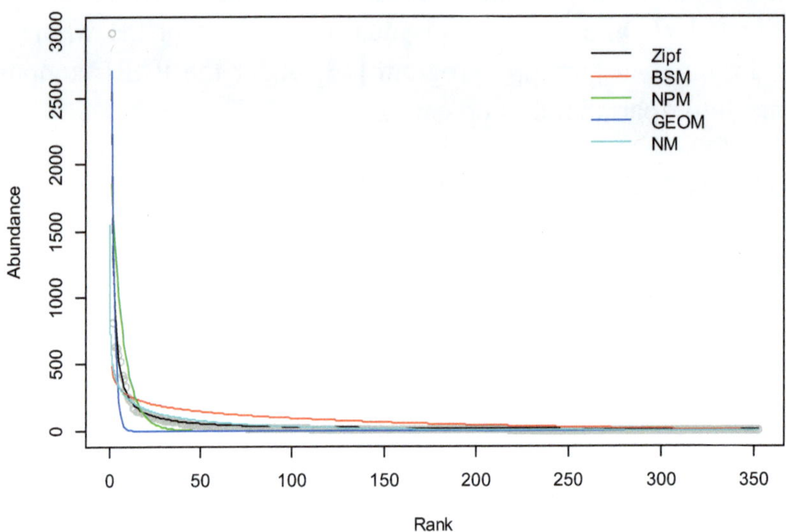

Figure 1: SAD for the whole microarthropod community and fitting of five statistical models. Codes: Zipf-Zipf model; BSM-broken stick model; NPM-niche preemption model; GEOM-geometric model; NM-neutral model.

Table 1: Evaluation of different models for fitting SADs of various microarthropod taxonomic groups. Codes: Zipf-Zipf model; BSM-broken stick model; NPM-niche preemption model; GEOM-geometric model; NM-neutral model. The best AIC model for each taxonomic group is marked in boldface. Asterisk denotes significant difference between expected and observed SAD with P<0.05. NA depicts non-applicable results without numeric information

Models	Evaluation Methods	Oribatids	Mesostigmatids	Collembolans	Whole Community
Zipf	AIC	424673.018	32225.320	4206503.747	987957.109
	χ^2 test	1543.139*	249.54*	2664.335*	1497.454*
	K-S test	0.689*	0.599*	0.759*	0.634*
BSM	AIC	1501178.146	595071.245	13403249.460	15965443.975
	χ^2 test	3198.674*	1496.289*	6070.837*	11212.600*
	K-S test	0.508*	0.569*	0.627*	0.577*
NPM	AIC	94857.987	205574.169	1387780.160	7052598.293
	χ^2 test	559.878*	617.622*	1551.823*	4535.233*
	K-S test	0.636*	0.803*	0.904*	0.858*

Table 1: contd...

GEOM	AIC	46259.356	95267.384	203697.032	3427744.618
	χ^2 test	622.263*	758.575*	1210.223*	5314.049*
	K-S test	0.689*	0.92*	0.94*	0.963*
NM	AIC	NA	NA	11724160	12601528
	χ^2 test	NA	NA	2274.287*	3076.281*
	K-S test	NA	NA	0.169	0.176*

SAD for Oribatids

As showed in Fig. **2** and Table **1**, all the models could fit the oribatid SAD quite well, but the difference between the expected and observed SAD still have a large significant discrepancy (indicated by Chi-square and K-S tests). Among the models, geometric model has the lowest AIC value, indicating the most favored model for oribatid SAD pattern.

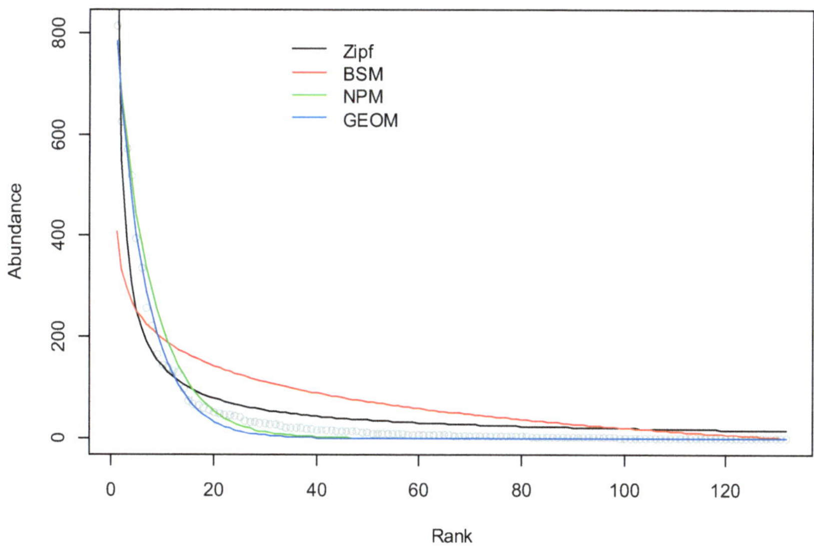

Figure 2: SAD for the orbatid community and fitting of four statistical models (neutral model (NM) failed to fit the data with odd parameter value θ=1.79855e+10, thus not being presented here). Codes: Zipf-Zipf model; BSM-broken stick model; NPM-niche preemption model; GEOM-geometric model.

SAD for Mesostigmatids

As showed in Fig. **3** and Table **1**, all the models could fit the mesostigmatid SAD quite well, but the difference between the expected and observed SAD still have a large significant discrepancy (indicated by Chi-square and K-S tests). Among the models, Zipf model has the lowest AIC value, indicating the most favored model for mesostigmatid SAD pattern.

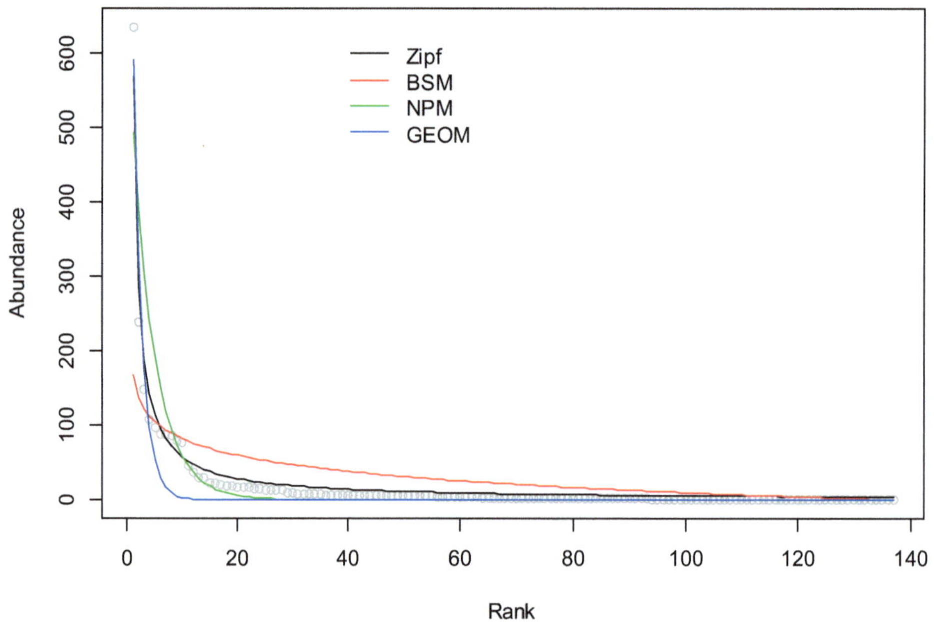

Figure 3: SAD for the mesostigmatid community and fitting of four statistical models (neutral model (NM) failed to fit the data with odd parameter value θ=2.47436e+10, thus not being presented here). Codes: Zipf-Zipf model; BSM-broken stick model; NPM-niche preemption model; GEOM-geometric model.

SAD for Collembolans

As showed in Fig. **4** and Table **1**, all the models could fit the collembolan SAD quite well, but the difference between the expected and observed SAD still has a large significant discrepancy (indicated by Chi-square and K-S tests). Among the models, geometric model has the lowest AIC value, indicating the most favored model for collembolan SAD pattern.

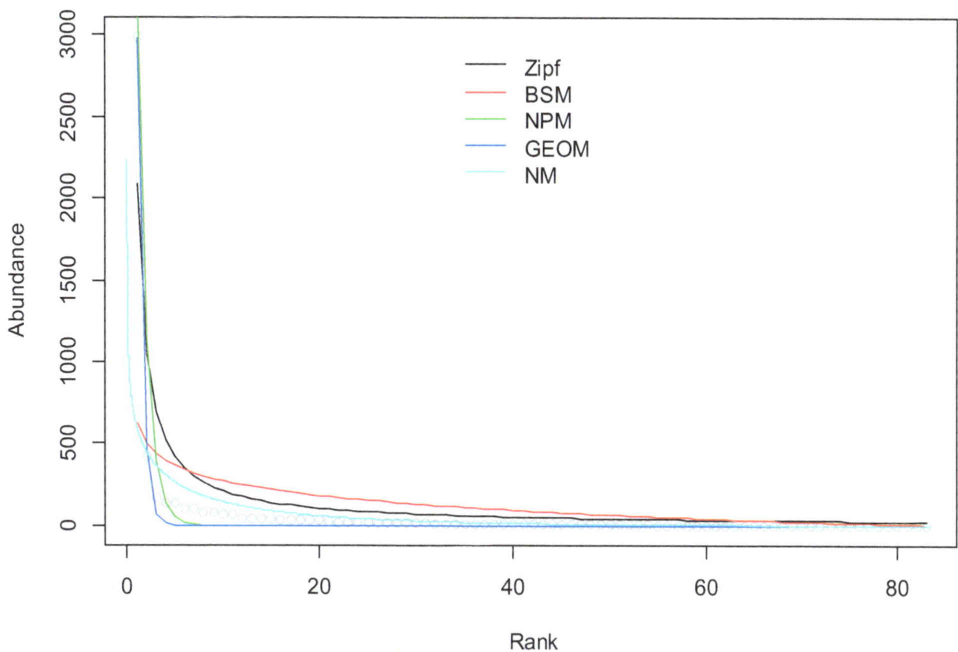

Figure 4: SAD for the collembolan community and fitting of five statistical models. Codes: Zipf-Zipf model; BSM-broken stick model; NPM-niche preemption model; GEOM-geometric model; NM-neutral model.

DISCUSSION

Previous studies have showed the importance of spatial scales [8] on influencing the selection of the best-fit model. At local scales, typically niche-derived models were found to have highest powers for plant communities [8, 11, 13]. In contrast, at large spatial scales, Hubbell's neutral model was found to be of highest power in many cases [8].

For microarthropod communities, I would originally expect that sampling area of my study (130 km×60 km) is large enough for microarthropod species. As a consequence, neutrality might be prevailing to influence SAD patterns. However, based on the results (Table **1**), it is broadly supported that niche-based Zipf and geometric models are the best models over different taxonomic groups based on AIC standard. Thus, my hypotheses were falsified.

One important reason for the contradictory predicted and fitted patterns above should be related to the mechanisms structuring SAD. Many studies have showed that it is not safe to draw relevant conclusions on the importance of niche and neutrality processes based on the fitting ability of exclusive models on SAD patterns. As mentioned earlier in the introduction, different mechanisms could result into similar SAD patterns [8, 22], and therefore the corresponding mechanisms could not be inferred.

My results are remarkably different from any previous empirical comparisons of different methods for fitting plant SADs [8, 12, 13] because I found either neutral or niche-relevant models could fit the microarthropod SADs models satisfying the requirement of statistical significance based on χ^2 and K-S tests. A key possible reason of the present observations might be because I fitted different SAD models on the raw data without any transformation. Typically, log-transformation is a common practice before fitting SAD models [23]. Thus, the present results might be altered to some extents if log-transformation of the raw abundance data of species is utilized.

REFERNECES

[1] J. Rosindell, S.P. Hubbell, F. He, L.J. Harmon, R.S. Etienne, "The case for ecological neutral theory.", *Trends Ecol. Evol.* 27, 203-208. doi:10.1016/j.tree.2012.01.004, 2012.

[2] J. Rosindell, S. Hubbell, R. Etienne, "The unified neutral theory of biodiversity and biogeography at age 10", *Trends Ecol. Evol.* 26, 340-348, 2011.

[3] S.P. Hubbell, The Unified Neutral Theory of Biodiversity and Biogeography (MPB-32) (Monographs in Population Biology), Princeton University Press, 2001.

[4] R. Ricklefs, S. Renner, "Global correlations in tropical tree species richness and abundance reject neutrality", *Science.* 335, 464-467, 2012.

[5] F. Munoz, P. Couteron, S. Hubbell, Comment on "Global correlations int ropical tree species richness and abundance reject neutrality," *Science.* 336, 1639, 2012.

[6] S. Nee, G. Stone, "The end of the beginning for neutral theory", *Trends Ecol. Evol.* 18, 433-434, 2003.

[7] S. Hubbell, "Neutral theory and the evolution of ecological equivalence", *Ecology.* 87, 1387-1398, 2006.

[8] J. Cheng, X. Mi, K. Ma, J. Zhang, "Responses of species-abundance distriubtion to varying sampling scales in a subtropical broad-leaved forest", *Biodivers. Sci.* 19, 168-177, 2011

[9] I. Volkov, J. Banavar, S. Hubbell, A. Maritan, "Neutral theory and relative species abundance in ecology", *Nature.* 424, 1035-1037, 2003.

[10] I. Volkov, J. Banavar, S. Hubbell, A. Maritan, "Patterns of relative species abundance in rainforests and coral reefs", *Nature.* 450, 45-49, 2007.

[11] L. Gao, R. Bi, M. Yan, "Species abundance distribution patterns of Pinus tabulaeformis forest in Huoshan Mountain of Shanxi Province, China", *Chin. J. Plant Ecol.* 35, 1256-1270, 2011.

[12] X. Du, S. Zhou, "Testing the neutral theory of plant communities in subalpine meadow", *J. Plant Ecol. Chin. Version.* 32, 347-354, 2008.

[13] Y. Yan, C. Zhang, X. Zhao, "Species-abundance distribution patterns at different succesional stages of conifer and broad-leaved mixed forest communities in Changbai Mountains, China", *Chin. J. Plant Ecol.* 36, 923-934, 2012.

[14] S. Walker, H. Cyr, "Testing the standard neutral model of biodiversity in lake communities", *Oikos*. 116, 143-155, 2007.

[15] S. Frontier, Diversity and structure in aquatic ecosystems, in: M. Rnes (Ed.), Oceanogr. Mar. Biol. Annu. Rev., Aberdeen University Press, Aberdeen, 1985: pp. 253-312.

[16] R. MacArthur, "On the relative abundance of bird species", *Pnas*. 43, 283-295, 1957.

[17] I. Motomura, "On the statistical treatment of communities", *Zool. Mag. Tokyo*. 44, 379-383, 1932.

[18] W. Bastow, "methods for fitting dominance diversity curves", *J. Veg. Sci*. 2 35-46, 1991.

[19] R.S. Etienne, "A new sampling formula for neutral biodiversity", *Ecol. Lett.* 8 (2005) 253-260. doi:10.1111/j.1461-0248.2004.00717.x.

[20] F. Jabot, R. Etienne, J. Chave, "Reconciling neutral community models and environmental filtering: theory and an empirical test", *Oikos*. 117, 1308-1320, 2008.

[21] Y. Chen, "A comparison on the impacts of short-term micro-environmental and long-term macro-climatic variability on structuring beta diversity of microarthrophod communities", *J. Asia-Pac. Entomol.* 17, 629-632, 2014.

[22] W. Harpole, D. Tilman, "Non-neutral patterns of species abundance in grassland communities", *Ecol. Lett.* 9, 15-23, 2006.

[23] B. McGill, R. Etienne, J. Gray, D. Alonso, M. Anderson, H. Benecha, *et al.*, "Species abundance distributions: moving beyond singla prediciton theories to integration within an ecological framework", *Ecol. Lett.* 10, 995-1015, 2007.

Chapter 10: Statistical Methods for Estimating Species Abundance

CHAPTER 10

Statistical Methods for Estimating Species Abundance

Abstract: As mentioned in the last chapter, species abundance and its probability distribution patterns were very basic and important units in ecological studies. Thus, for ecological studies, it is very natural and common to collect species abundance data. However, limited sampling of species abundance could not accurately reflect the abundance and density patterns of species in the sampling areas. Thus, it is necessary to utilize statistical methods to infer the abundance of specie with different experimental designs. In this chapter, I outlined the relevant statistical methods for estimating species abundance when using different sampling strategies.

Keywords: Adaptive sampling, biodiversity analyses, biodiversity survey and inventory, ecological sampling, greedy algorithm, random sampling, sampling strategies, species richness and abundance, statistical inference, statistical uncertainty, statistical variance.

INTRODUCTION

Species abundance is one of the most important metrics in ecological studies. The attempts for accurately estimating species abundance are not a new emerging subject [1]. There are many statistical methods proposed by mathematicians and ecologists to better estimate species abundance [1-3]. However, most of these methods are on the basis of random sampling and the distribution of species is spatially random and not aggregated. It is always violated for such an assumption in the real-world situations since species distribution usually presents aggregation patterns [4-6].

Adaptive cluster sampling (ACS) is a method for estimating species abundance for handling the situation of distributional aggregation [7, 8]. In comparison to traditional sampling methods, ACS requires iterative procedures to detect networks in which each plot has the number of individuals of the focused species above a threshold value (C). ACS is much more powerful for estimating species abundance when the distribution of the species is very rare or aggregated [9]. In recent years, evaluation of the efficiency of adaptive cluster sampling for estimating population size mean and variance across different taxa has been carried out [9, 10].

Youhua Chen

Before performing a series of simulation and analytical comparison, I have the following predictions pending to be tested: (1) the estimation of species abundance using RS sampling strategy tends to have larger variance because the sampled grids can be anywhere over the whole area. In contrast, ACS algorithm should be more efficient as evidenced in previous studies [10-13]. (2) Increasing sampling intensity (reflected by the number of sampling grids) will reduce the variance for the estimation of species abundance for both ACS and RS cases.

RANDOM SAMPLING ALGORITHM (RS)

The random sampling (RS) algorithm is to choose *n* grid cells randomly without replacement over the whole area and measure associated spatial point statistics and estimate the total abundance over the whole area as follows,

$$\hat{\tau}_{RS} = \frac{N}{n} \sum_{i \in n} \tau_i \tag{1}$$

where τ_i is the total number of individuals found in the *i*-th grid of the *n* sampling grids and *N* is the total number of grids of the whole area.

The variance of the estimator is given by,

$$Var(\hat{\tau}_{RS}) = \frac{N^2}{n(n-1)} \left(1 - \frac{n}{N}\right) \sum_{i \in n} (\tau_i - \overline{\tau})^2 \tag{2}$$

where τ_i is the mean individual number for the *n* sampling grids.

ADAPTIVE CLUSTER SAMPLING ALGORITHM (ACS)

Here, a network is defined as a group of grid cells that are adjacent from each other, which is identified by ACS algorithm. The number of individuals for each of the neighboring cells adjacent to the edges of a network should be less than *C*.

ACS algorithm is implemented as follows: it starts from some grid cells which are randomly selected. Each of these cells constitutes an initial network. For each initial network, ACS algorithm stops to search neighboring cells when the number of individuals of the species in the network less than the threshold value *C*. If an initial network satisfies the threshold condition, ACS algorithm will continue searching the neighboring cells of the initial network. Those the neighboring cells with species individual number larger than or equal to the threshold value *C* will become part of the network. As such, the network will grow bigger as the algorithm goes forward. The algorithm will stop to find neighboring cells for a network until that all the neighboring cells of the network fail to satisfy the requirement of threshold condition. For simplicity, in the present study, we always set C=1. That is, a neighboring grid cell will be merged into a network if there is at least one species individual inside. The ACS algorithm terminates when the given number of sampling grid cells is arrived.

Hansen-Hurwitz (HH) estimator [7, 9, 10, 14] is used to estimate total abundance. When the initial random sample of n grid cells is selected with replacement, for HH estimator, mean population number for each grid cell is given by,

$$\hat{\mu}_1 = \frac{1}{n} \times \sum_{i=1}^{n} w_i = \frac{1}{n} \times \sum_{i=1}^{n} \frac{\tau_i}{m_i} \tag{3}$$

where n is the number of networks using ACS algorithm. τ_i is the number of individuals in the i-th network while m_i the number of grid cells that the i-th network contains.

The corresponding variance for the mean population per grid cell is given by,

$$Var(\hat{\mu}_1) = \frac{1}{n(n-1)} \times \sum_{i=1}^{n} \left(\frac{\tau_i}{m_i} - \hat{\mu}_1\right)^2 \tag{4}$$

The estimated total number of species individuals over the whole area reads,

$$\hat{\tau}_{HH} = N\hat{\mu}_1 = \frac{N}{n} \times \sum_{i=1}^{n} \frac{\tau_i}{m_i} \tag{5}$$

where N is the total grid cells for the whole area.

The variance is,

$$Var(\tau_{HH}) = N^2 Var(\hat{\mu}_1) \tag{6}$$

when initial random sample of n grid cells is selected without relacement, the variance of the population per grid cell is different from equation (2), while the formula for the mean is the same as (1):

$$Var(\hat{\mu}_1) = \left(\frac{N-n}{N}\right) \times \frac{1}{n(n-1)} \times \sum_{i=1}^{n} (\frac{\tau_i}{m_i} - \hat{\mu}_1)^2 \quad (7)$$

In the present study, the ACS algorithm is implemented without replacement on the initial choice of the grid cells. Thus, the variance obtained from equation (7) will be presented only in the result section.

All the above-mentioned statistical analyses are done under R environment [15]. The R scripts for implementing ACS sampling strategy are available upon request from the author. When evaluating the efficiency of RS and ACS sampling strategies on recovering spatial point patterns of full observed data points, the following sampling sizes are considered: 5, 10, 30, 60 and 100. As expected, the likelihood of detecting actual spatial point patterns and total abundance from full observed data would increase when the sampling sizes are increased. For each sampling size, 500 replicates are applied to guarantee consistence. All the above statistical methods are re-run for each replicate. Estimated abundance is calculated for each sampling size using the overall average value over the 500 replicates.

When using RS or ACS sampling methods, it is very likely that the grid cells randomly obtained may not contain any distributional points of species. In this situation, no spatial point metrics and inference of spatial point processes can be done. Thus, when it happens, the failure of the sampling is counted cumulatively. This count of sampling failure may be a way to measure the efficiency of RS and ACS sampling methods.

A PRACTICAL EXAMPLE

The distribution of four Soricomorpha species in Poland was collected from a previous study [16]. For performing systematic sampling, the grid system with a resolution of $0.5^\circ \times 0.5^\circ$ (latitude times longitude) is considered for covering the whole area of Poland (Fig. **1**).

RESULTS AND DISCUSSION

As seen in Tables **1** and **2**, when the sampling size is small (less than 10), ACS is more accurate than RS algorithm for estimating species abundance since the results from ACS algorithm have smaller standard deviation and the upper bound of species abundance (mean + standard deviation) are usually larger than the true abundance. In contrast, the estimated abundance derived from RS algorithm usually has larger variance in comparison to those for ACS algorithm.

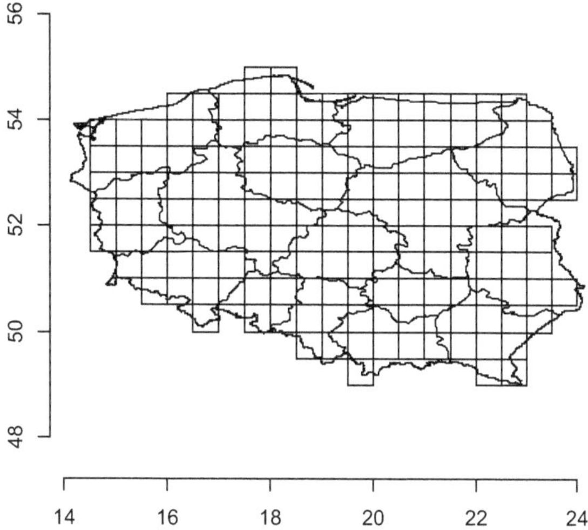

Figure 1: The $0.5^{\circ} \times 0.5^{\circ}$ Poland grid system used for performing quadrat sampling and the estimation of species abundance. Numbers in the X- and Y-axis denote the latitude and longitude in degrees.

Interestingly, when the initial sampling size increased over 30 (Tables **1** and **2**), both algorithms worked unsatisfactorily since the estimated abundance for the four Soricomorpha species were very different from the true values.

Similar to previous studies, both RS and ACS methods are sensitive to the number of initial grid cells (sampling size here) (Tables **1** and **2**). If more initial grid cells are selected, the estimated abundance will have a trend to reduce variance for both ACS and RS cases.

Table 1: Observed and estimated abundance of each of the four species using RS sampling strategy with 500 replicates. Values before the sign " ± " depict the estimated mean of abundance while values after the sign " ± " denote standard deviation of abundance. All the mean and standard deviation values are from the average values from the 500 replicates. Values inside the parentheses in the first column denote the observed individual numbers (true abundance) for the whole area for the species

Species	**Sample number**				
	5	**10**	**30**	**60**	**100**
Talpa europaea (85)	3.8 ± 63.7	2.9 ± 54.5	1.1 ± 28.7	0.4 ± 16.9	0.3 ± 11.7
Sorex araneus (210)	31.5 ± 293.8	18.0 ± 218.2	6.3 ± 111.9	3.1 ± 72.8	1.9 ± 45.7
Sorex minutus (148)	19 ± 204.9	13.4 ± 164.8	4.3 ± 83.7	2.1 ± 52.9	1.3 ± 34.1
Neomys fodiens (31)	10.7 ± 133.1	7.7 ± 109.9	2.5 ± 57.4	1.2 ± 57.4	0.8 ± 23.5

Table 2: Observed and estimated abundance of each of the four species using ACS sampling strategy with 500 replicates. Values before the sign "± " depict the estimated mean of abundance while values after the sign "±" denote standard deviation of abundance. All the mean and standard deviation values are from the average values from the 500 replicates. Values inside the parentheses in the first column denote the observed individual numbers (true abundance) for the whole area for the species

Species	Sample number				
	5	10	30	60	100
Talpa europaea (85)	27 ± 60.6	33.4 ± 47.6	26.6 ± 23.1	24.3 ± 14.3	26.3 ± 8.6
Sorex araneus (210)	103.2 ± 157.3	112.6 ± 114.2	38.4 ± 31.4	39.2 ± 20.1	38.7 ± 12.0
Sorex minutus (148)	87.5 ± 156.0	98.2 ± 102.8	72.7 ± 44.7	70.9 ± 28.1	73.5 ± 17.1
Neomys fodiens (31)	55.2 ± 104.9	71.1 ± 76.5	51.4 ± 35.5	50.8 ± 21.9	53.0 ± 13.4

For estimating species abundance, when the initial sampling size increased over 30, both ACS and RS algorithms worked unsatisfactorily. There may be two major reasons that can explain this phenomenon. Firstly, when the number of sampling size is increased, more grids with zero individuals will be included. As such, when using the equations (1) and (3) to estimate the density of species' abundance ($\hat{\tau}_i$), the smaller values would be returned. The smaller density values will lead to the estimation of species abundance become much smaller ($\hat{\tau}_{RS}$ and $\hat{\tau}_{HH}$). Second, for ACS algorithm, because the total grid cells for the whole area (Poland here) is fixed, when the initial sample size is increased, the likelihood that two or more networks are overlapped when searching neighbours of the network will increase. Given that the whole Poland area has 165 grid cells at $0.5^\circ \times 0.5^\circ$ spatial grain, when the sampling size is increased to 100. It is very easy that some of the 100 networks would overlap in some of their grid cells. As consequences, utilization of equations (1) and (3) would have high risks to re-count grid cells in the overlapping zones and associated individual numbers multiple times, and thus cause biases. Moreover, the usage of equation (7) to calculate variance would be improper when sampling size is too large (when overlapping of networks happen, equation (6) should be more rationale).

The present chapter might have some ecological implications because in reality, the distribution information of species over the whole landscape and region is usually not available but only available for some sampling areas, which just account for a limited fraction of the whole concerned region.

REFERENCES

[1] G. Seber, Estimation of animal abundance, Blackburn Press,Caldwell, New Jersey, USA, 2002.

[2] F. He, K. Gaston, "Estimating abundance from occurrence: an underdetermined problem", *Am. Nat.* 170, 655-659, 2007.

[3] F. He, K. Gaston, "Estimating species abundance form occurrence", *Am. Nat.* 156, 553-559, 2000.

[4] L. Taylor, Aggregation, "variance and mean", *Nature*. 189, 732-735, 1961.

[5] M. Gao, "Detecting spatial aggregation from distance samplign: a probability distribution model of nearest neighbor distance", *Ecol. Res.* 28, 397-405, 2013.

[6] C. Hui, C. Boonzaaier, L. Boyero, "Estimating changes in species abundance from occupancy and aggregation", *Basic Appl. Ecol.* 13, 169-177, 2012.

[7] S. Thompson, "Adaptive cluster sampling", *J. Am. Stat. Assoc.* 85, 1050-1059, 1990.

[8] S. Thompson, G. Seber, Adaptive sampling, John Wiley & Sons, Ltd, New York, USA, 1996.

[9] P. Ojiambo, H. Scherm, "Efficiency of adaptive cluster sampling for estimating plant disease incidence", *Phytopathology*. 100, 663-670, 2010.

[11] N. Goldberg, J. Heine, J. Brown, "The application of adaptive cluster sampling for rare subtidal macroalgae", *Mar. Biol.* 151, 1343-1348, 2007.

[12] B. Acharya, G. Bhattarai, A. de Gier, A. Stein, "Systematic adaptive cluster sampling for the assessment of rare tree species in Nepal", For. *Ecol. Manag.* 37, 65-73, 2000.

[13] M. Arabkhedri, F. Lai, I. Noor-Akma, M. Mohamad-Roslan, "An application of adaptive cluster sampling for estimating total suspended sediment load", *Hydrol. Res.* 41, 63-72, 2010.

[14] M. Salehi, G. Seber, "Adaptive cluster sampling with networks selected without replacement", *Biometrika*. 84, 209-219, 1997.

[15] R Development Core Team, R: A Language and Environment for Statistical Computing, Vienna, Austria. ISBN 3-900051-07-0, URL http://www.R-project.org., (2013).

[16] A. Banaszek, W. Chetnicki, S. Fedyk, K. Jadwszczak, P. Mirski, "Data on the distribution of selected soricomorpha and rodentia species in Poland", *Zool. Pol.* 57, 65-103, 2012.

Chapter 11: Testing Distributional Randomness

CHAPTER 11

Testing Distributional Randomness

Abstract: Species distribution is reflected by the observation of species individuals in different sites of an ecological domain. The distributional size and the range boundary of the species are measured through the counting of the presence or absence of the species across different sites. However, as mentioned in previous chapters 9 and 10 and here, when the individuals of species has been recorded with the geographic coordinates within some spatial ranges, one common ecological question that we would encounter would be, are the distribution of all the species individuals present some non-random patterns? That is, would the species tend to aggregately distribute in some limited areas across the whole ecological domain, while some areas are not favoured by the species and thus rarely have the occurrence of the individuals of species? Testing of distributional randomness could offer insights and address these ecological questions. In this chapter, I would present only some classical statistical methods for quantifying the randomness of spatial distribution of species. Readers interested into this subject should refer to books in the field of spatial statistics.

Keywords: Distribution and diversity, limited sampling, nearest neighbouring, quadrate-based versus spatial point-based randomness, Riley's K statistic, spatial biodiversity patterns, spatial statistics.

INTRODUCTION

There are three basic species distribution patterns: random, aggregate and regular (Fig. **1**). Random distribution does not present any trends or spatial aggregation, while aggregate distribution presented some local aggregation of species' distributional points. As seen in the figure, there are two local clusters on the distribution of the species, one is located in the left-upper side of the sampling layer while another cluster/aggregation center is located in the right-lower part of the layer. At last, regular distribution is another extreme situation, in which the distributional points of species presented very regular patterns and the distances among the pairs of the points presented some constant values (Fig. **1**).

Random distribution may be the Poisson distribution when the mean number of points per unit area equals the variance of distances between the points. Aggregate distribution occurs when individuals of a species tend to be close to each other. Finally, a regular distribution occurs when individuals of a species tend to avoid one another and thus their distances are as large as possible. Consequently, each of the spatial points of a regular distribution is usually located at a fixed distance from its neighbors. In statistical sense, an aggregated, regular or random

Youhua Chen

distribution is one in which the variance/mean ratio of the number of distributional points in sampled quadrats is significantly larger, lower or equal to one [1].

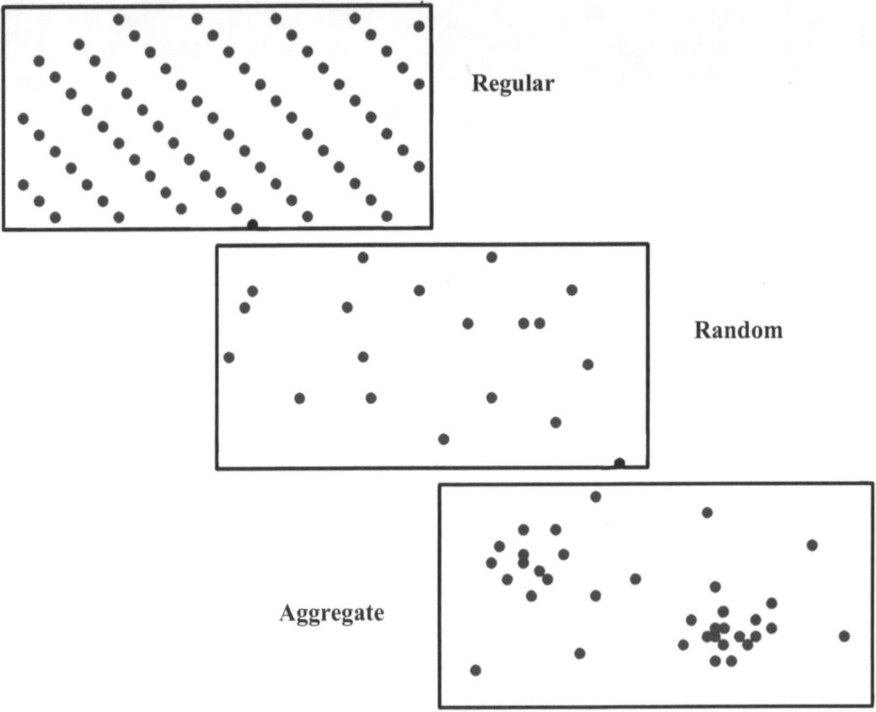

Figure 1: A hypothetical example for demonstrating the regular, random and aggregate distribution of species in a hypothetical sampling area in a rectangle.

Quadrat-Based Method-Measurement of Distributional Aggregation of Species

As mentioned in some literature [2, 3], the Poisson distribution and negative binomial distribution [4-7] was one of the most widely used statistical probabilities to quantify the distributional randomness of species.

The Poisson distribution is written as,

$$P(n) = \frac{\lambda^n e^{-\lambda}}{n!}, n = 0,1,2,... \quad (1)$$

where λ is the mean individuals of species found in a cell/quadrat, n is the number of the individuals for a species found in a sampling layer.

The negative binomial distribution is given by,

$$P(n) = \binom{n+k-1}{n} \left(\frac{\mu}{\mu+k}\right)^n \left(\frac{k}{\mu+k}\right)^k, n = 0,1,2,... \tag{2}$$

where k here is the aggregation coefficient. Small k indicate high aggregation degree of species' distribution, while large k indicate low aggregation of species' distribution.

We now introduced the finite version of negative binomial distribution [2, 3] to measure distributional aggregation of species.

The finite negative binomial probability of species distribution reads,

$$P_{FNBD}(n) = \frac{\binom{n+k-1}{n}\binom{N-n+k/a-k-1}{N-n}}{\binom{N+k/a-1}{N}} \tag{3}$$

estimation of aggregation parameter k follows the previous work [3]:

$$\begin{cases} \hat{k} = \dfrac{(1-a)\overline{n}^2 - as^2}{s^2 - (1-a)\overline{n}} \\ \overline{n} = \dfrac{1}{m}\sum n_i \\ s^2 = \dfrac{1}{m}\sum (n_i - \overline{n})^2 \end{cases} \tag{4}$$

where m is the number of quadrats, a is the ratio between smallest spatial unit size and the whole range, $\{n_i\}$ is the abundance vector in a set of sampled quadrats. A high (or low) k value indicates that the species will have a low (or high) aggregation on its spatial distributional pattern.

Spatial Point-Based Method-Measurement of Distributional Aggregation of Species

Spatial point patterns gain growing attention in recent years in the field of ecology [8-10]. The distributional patterns of species can be well elucidated and associated ecological mechanisms can be successfully quantified using spatial point analysis [11, 12]. Spatial point pattern analysis is quite commonly used in plant ecology.

Here, we introduce Riley's *K* functions[13-17] for quantifying distributional randomness as follows.

The intensity of the spatial points λ is one of the most fundamental metrics, which is simply calculated as the ratio between total observed point number *N* and sampling area size *A*:

$$\lambda = \frac{N}{A} \tag{5}$$

Ripley's *K* function is defined as,

$$K(s) = \frac{1}{\lambda} \sum_{\substack{i,j \\ j \neq i}} w_{ij} I(d_{ij} < s) / n \tag{6}$$

Where *n* is the number of total observed distributional points, $I(\bullet)$ is the indicator function; d_{ij} is the distance between *i*-th and *j*-th points. When distance d_{ij} is smaller than given distance *s*, $I(d_{ij} < s)=1$; λ is the density of the observed points over the sampling area. w_{ij} is the edge correction weight. Under complete spatial randomness (CSR), the expected *K* function is $K(s) = \pi s^2$ (5).

Since *K(s)* is a function of distance *s*, the results are presented in terms of graphs. We consider the integral of *K(s)* observed and null values under CSR for each sampling over the distance *s* to compare observed and expected *K* values generated from the homogeneous Poisson process and thus determine whether the spatial point pattern is a cluster, regular or random one:

$$\hat{K} = \int_{s=0}^{l} K(s) ds \tag{7}$$

Where *l* is the maximum distance that *s* can reach when the *K(s)* values are computed. Under CSR, the overall integral of *K(s)* is given by,

$$\hat{K}_{CSR} = \frac{2\pi l^3}{3} \tag{8}$$

$\hat{K}_{obs} > 2\times\hat{K}_{CSR}$ suggests an aggregation pattern, $\hat{K}_{obs} < \hat{K}_{CSR}/2$ a regular pattern, and $\hat{K}_{CSR}/2 \le \hat{K}_{obs} \le 2\times\hat{K}_{CSR}$, a random pattern of the points. This criterion is a bit arbitrary, but it is inspired from the calculation of a 95% confidence interval (where a standard deviation is multiplied by a value of 1.96).

The L function is also widely used, which is related to Ripley's K function as follows,

$$L(s) = \sqrt{K(s)/\pi} \tag{9}$$

If the distribution of species is homogeneous or fully random, then the plot of $s - L(s)$ *versus* s should follow the horizontal zero-axis.

Another function, called G function or the nearest neighbour distribution function, is defined as the probability of the distributional points of a species of which the distances to the nearest other points of the species is less than or equal to a given distance s.

When species' distribution is homogeneous and stationary, the corresponding null model for the G function is given by,

$$G(s) = 1 - \exp(-\pi\lambda s^2) \tag{10}$$

where λ denotes the intensity of the spatial points as mentioned above.

REMARKS

In spatial statistics, these statistics could be applied to higher dimensional spaces. Their definitions are similar, but some additional notations should be given.

REFERENCES

[1] K. Wilson, O. Bjornstad, A. Dobson, S. Merler, G. Poglayen, S. Randolph, *et al.*, "Heterogeneities in macroparasite infecitons: patterns and processes", in: *Ecol. Wildl. Dis.*, Oxford University Press, Oxford, 2002: pp. 6-44.

[2] Y. Chen, "A multiscale variation partitioning procedure for assessing the influence of dispersal limitation on species rarity and distribution aggregation in the 50-Ha tree plots of Barro Colorado Island, Panama", *J. Ecosyst. Ecography*. 3, 134, 2013.

[3] T. Zillio, F. He, "Modeling spatial aggregation of finite populations", *Ecology*. 91, 3698-3706, 2010.

[4] E. Pielou, Mathematical Ecology, Wiley, New York, 1977.

[5] L. Taylor, I. Woiwod, J. Perry, "The density dependence of spatial behaviour and the rarity of randomness", *J. Anim. Ecol.* 47, 383-406, 1978.

[6] J. Perry, L. Taylor, Ades: "new ecological families of species-specific frequency distributions that describe repreated spatial samples with an intrinsic power-law variance-mean property", *J. Anim. Ecol.* 54, 931-953, 1985.

[7] M. Boswell, G. Patil, "Chance mechanisms generating the negative binomial distribution", in: G. Patil (Ed.), Random Counts Models Struct., Pennsylvania State University Press, University Park, Pennsylvania, USA, 1970: pp. 3-22.

[8] T. Wiegand, K. Moloney, "Rings, circles and null-models for point pattern anlaysis in ecology", *Oikos*. 104, 209-229, 2004.

[9] S. Getzin, C. Dean, F. He, T. Trofymow, K. Wiegand, T. Wiegand, "Spatial patterns and competition of tree species in a Douglas-fir chronosequence on Vancouver Island", *Ecography*. 29, 671-682, 2006.

[10] T. Wiegand, W. Kissling, P. Cipriotti, M. Aguiar, "Extending point pattern analysis to objects of finite size and irregular shape", *J. Ecol.* 94, 825-837, 2006.

[11] P. Diggle, Statistical analysis of spatial point patterns, Academic Press, London, UK, 2003.

[12] S.M. Eckel, "Statistical Analysis of Spatial Point Patterns: Applications to Economical, Biomedical and Ecological Data", Universität Ulm 2008, 2008.

[13] A. Baddeley, J. Moller, R. Waagepetersen, "Non- and semi-parametric estimation of interaciton in inhomogeneous point patterns", *Stat. Neerlandica*. 54, 329-350, 2000.

[14] E. Marcon, F. Puech, "Measures of the geographic concentraiton of industries: improving distance-based methods", J. Econ. Geogr. 10, 745-762, 2010.

[15] B. Ripley, "Modelling spatial patterns", *J. R. Stat. Soc.* 39, 172-212, 1977.

[16] B. Ripley, "The second-order analysis of stationary point processes", *J. Appl. Probab.* 13, 255-266, 1976.

[17] B. Ripley, Statistical inference for spatial processes, Cambridge University Press, Cambridge, UK, 1988.

CHAPTER 12

Modeling Species' Potential Distribution

Abstract: Species distribution modeling (SDMs) may be one of the most exciting and hottest topics in current ecological studies. SDMs have been widely employed for ecologists to evaluate how climate change may drive species to go extinction under future climate conditions. SDMs are also used to monitor and delineate the cryptic and newly described species that have little information about their biology and distribution. Overall, SDMs are powerful tools for conducting biodiversity analyses. In this chapter, I outlined some of the most commonly used algorithms for conducting SDMs. A practical example is also provided.

Keywords: Black box issue, climate change, ecological niche analysis, ensemble species distribution modeling, fundamental niches, geographical information system, global change, land use, macroecology, model training and test, niche conservatism hypothesis, realized niches, species distribution, species extinction, statistical algorithms, statistical ecology.

INTRODUCTION

Biodiversity conservation [1] is one of major tasks for ecologists as repeatedly stated previously. There are two principal and basic units when doing conservation prioritization analyses as illustrated in previous chapters: area and species. Thus, identifying the suitable area for species to survive and fight against climate change would be important to guarantee successes in a species conservation program. Potentially suitable habitat prediction has become an effective means in wildlife conservation [2-4].

Species' potential suitable areas prediction heavily relies on statistical algorithms and geographic information systems as the results of the prediction are the suitable range maps for species [4]. The statistical methods and algorithms for doing the prediction of species suitable range is to use presence and absence information of species across the sampling sites as the input [5, 6]. In most cases, absence information of species is not available, presence-only statistical methods for predicting species' suitable habitat range are widely applied and tested for their predictive powers. In this chapter, I outlined some of the widely used statistical algorithms for predicting species' potentially suitable habitat ranges so as to allow the readers to become familiar with the techniques and advances in modeling species' potential suitable habitat.

Youhua Chen

STATISTICAL ALGORITHMS

GARP

The GARP method, the abbreviation of the full name 'Genetic Algorithm for Rule-set Production', was used as the ecological niche model to predict the potential distribution. As one of genetic algorithms, GARP model works in an iterative manner for rule construction, evaluation and incorporation or rejection to produce a heterogeneous rule-set describing the species' ecological niche (Stockwell & Peters 1999; Peterson & Vieglais 2001; Zhu *et al.,* 2007). As a genetic algorithm, GARP model has been applied widely in modeling species potential distribution (Peterson & Vieglais, 2001; Peterson, 2005; Ron, 2005; Li *et al.,* 2006; Zhu *et al.,* 2007). Herein we briefly introduce the genetic algorithms, which develop rules to build niche models by a process analogous to natural selection.

Firstly, the performance of the set of generated rules is evaluated. Only those rules that gained the highest performance were utilized for next-step simulation. Then, genetic processes including mutation, recombination, and crossing are generated for the purpose to randomly modify the rules identified previously with the highest scores. Such a genetic-like simulation process is repeated until that the additional generations do not improve the performance of those rules based on the model evaluation (Stockwell & Peters, 1999; Ron, 2005).

MAXENT

MaxEnt is a maximum entropy algorithm [7-10] that estimates the probability distribution for a species' occurrence based on the actual occurrence points and the defined environmental constraints. In principle, maximum entropy model is to maximize the following quantity which quantifies the entropy of the system:

$$\max \sum_i p_i \ln p_i$$

This maximization procedure is subjected to some sorts of constraints that are related to environmental conditions of the species occurrence. Here p_i is the occurrence probability of the modeled species in the site *i*.

BIOCLIM

The idea of BIOCLIM [11, 12] is to identify all areas with a similar climate to the locations where the species is found to present before the modeling based on some sorts of distance metrics (*e.g.*, Euclidean distance metrics).

Ecological-Niche Factor Analysis (ENFA)

Ecological-Niche Factor Analysis (hereafter ENFA) is a presence-only modeling technique, which can be used to identify the correlations between ecogeographical variables (EGVs) and species distribution. ENFA is to compare the distribution of the EGVs between the presence information of species and the whole study area. The outputs of the ENFA include factor scores and eigenvalues.

There are two important indices for measuring the ENFA outputs: the 'marginality factor' is defined as the standardized difference between the species mean and the global mean of all descriptors. The coefficients of the scores matrix related to the marginality factor indicate the correlation between each EGV and species' distribution. Another important index is 'specialization factors', quantifying how specialized the species is with respect to each EGV. Higher coefficients are associated with a more restricted value of each EGV [3, 13].

A PRACTICAL EXAMPLE

Suitable Range Prediction of the Haplotypes of *Chrysanthemum Indicum* in China Using Ecological Niche Modeling [14]

Distribution and Haplotype Information

I used the distributional records and haplotype information of *Chrysanthemum indicum* published by a previous study [15], in which 27 populations have been geo-referenced. In their study, 15 haplotypes were recorded, but I could not find any information about haplotype 6 and its distribution over populations. As such, I discarded this haplotype and the remaining 14 haplotypes were used in the subsequent analyses. The site (or populations) and haplotype data set is therefore a 27×14 matrix, which is subjected to the analysis of variation partitioning.

Spatial and Environmental Variables

The latitude and longitude are the two most important spatial variables being used as the spatial variables. However, because the raw spatial variables might not be accurate to characterize the structure of spatial autocorrelation inside the haplotype data, I performed a transformation using principal coordinates of neighboring matrices (PCNM) [16, 17].

Fifteen environmental variables were gathered from the IPCC (http://www.ipcc.ch) and Hydro 1k (http://webgis.wr.usgs.gov/globalgis/metadata_qr/metadata/hydro1k.htm) dataset, including climatic and physical ones. They were temperature range, annual frost, (average monthly) precipitation, solar radiation, annual minimum temperature, annual mean temperature, maximum temperature, annual evaporation, annual humidity, elevation, landscape, slope, irrigation direction, irrigation accumulation, and vegetation [18]. These data are rescaled into a unified spatial resolution of $0.25^{\circ} \times 0.25^{\circ}$. The mean and standard deviation (sd) of each of these variables over the grids of the territory of China, including another two spatial variables (latitude and longitude), are summarized in Table **1**.

Table 1: Summary statistics of environmental and spatial variables over grids of China used for the present study (mean+/-standard deviation)

Longitude ($^{\circ}$)	103.95+/-14.37	Vegetation index	127.00+/-74.90
Latitude ($^{\circ}$)	36.34+/-7.18	Temperature range ($^{\circ}C$)	9.26+/-7.29
Annual frost (day)	114.4+/-63.71	Precipitation (mm)	15.35+/-12.91
Landscape index	120.09+/-76.50	Radiation (W / m^2)	147.48+/-14.57
Elevation (m)	117.82+/-74.50	Minimum temperature ($^{\circ}C$)	3.74+/-53.46
Irrigation accumulation	13.67+/-37.59	Mean temperature ($^{\circ}C$)	7.21+/-6.61
Irrigation direction	30.53+/-39.24	Maximum temperature ($^{\circ}C$)	12.06+/-6.84
Slope ($^{\circ}$)	92.60+/-73.91	Annual evaporation (mm)	7.47+/-5.77
Relative annual humidity	8.61+/-4.06		

To avoid potential collinearity among the environmental variables when doing variation partitioning, I performed a principal component analysis and extracted the first five principal axes as the representative of important environmental gradients influencing population structure of *C. indicum*.

Model Configuration

I used one of most robust methods, maximum entropy method, to predict the suitable ranges of species using the software Maxent [9, 10, 19]. The abovementioned fifteen variables and 27 presence records were used for training and testing. Default setting in Maxent and the logistic outputs were adopted. In specific, 500 maximum iterations were utilized, convergence threshold was set to 0.00001. Random test point percentage was 25%. 15388 grid cells were used as the background and presence points when modeling the distribution for *C. indicum*. We also evaluated the most correlated environmental variables accounted for predicting the distribution of the species using alternative methods using the output of Maxent (percent contribution, permutation importance and jackknife test). Hinge features were applied. Model evaluation was carried out using areas under the curve (AUC) and maximum true skill statistic (max_TSS) indices [20-23]. All the data layers were projected onto geographical maps with a resolution of 10 minutes [23]. The mapping was confined to the region of East and South East Asia.

Results and Discussion

The Maxent algorithm was converged at the time step of 480. As indicated by Fig. **1**, the most suitable ranges were mainly situated on the eastern and central part of China. However, some trivial predicted distributional ranges could be found in the southwestern and southern part of China as well. The prediction is robust since the model evaluation indices AUC and max_TSS are very high for the training data (0.953 and 0.940 respectively).

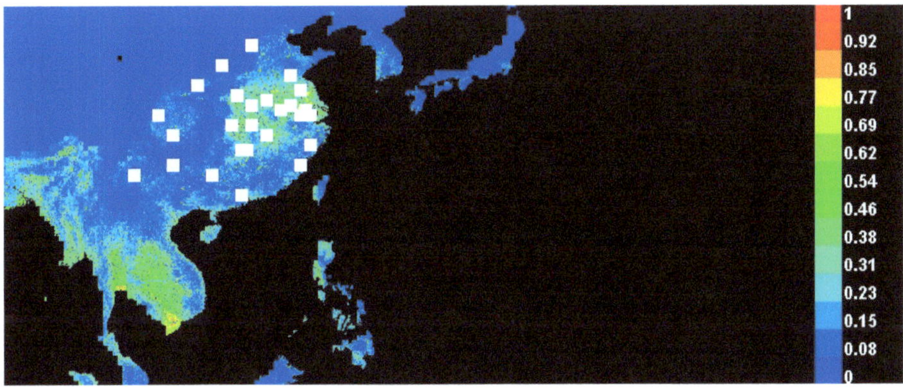

Figure 1: Suitable ranges of *C. indicum* in S Asia. Colors from blue to red indicated the potential occurrence probability from low (0) to high (1). White squares indicated the distributional records of *C. indicum* populations used in the present study.

The most influential variable measured by percent contribution (Table **2**) indicated that minimum temperature played a dominant role in modeling species potential distribution, followed by maximum temperature and annual humidity. Permutation improtance basically supported the importance of these three variables, but the order is different (Table **2**).

Furthermore, the jackknife test of regularized training gain (Fig. **2**) evidenced the importance of minimum temperature (indicated by the length of blue bar). However, this time annual frost became the most important variable, which was identified to have no influence in percent contribution. Jackknife test suggested that it contained the largest amount of information to predict the potential range of *C. indicum*. In contrast, another variable annual humidity contained most special/unique information that other variables did not hold to predict the potential range of *C. indicum*. This variable was ranked as one of the important top 3 variables in the measures of percent contribution and permutation importance (Table **2**).

Table 2: Percent contribution and permutation importance of explanatory variables in the modeling of suitable ranges of *C. indicum*. Percent contribution reports the gain of the model by including a particular variable at each step of the Maxent algorithm, while permutation importance reports the contribution for each variable to the final Maxent model which is determined by randomly permuting the values of that variable among the training points and measuring the resulting decrease in training AUC

Variable	Percent contribution	Permutation importance
Minimum temperature	63.3	9.3
Maximum temperature	8.9	52.3
Annual humidity	6.7	16.9
Vegetation	4.8	5.6
Precipitation	4.5	0.4
Slope	3.3	0.9
Irrigation accumulation	2.6	3.6
Landscape	2.5	3.3
Mean temperatuer	1.3	0
Annual evaporation	1.2	7.3
Elevation	0.8	0.1
Temperature range	0	0.3
Solar radiation	0	0
Irrigation direction	0	0
Annual frost	0	0

Therefore, all the three metrics consistently identified minimum temperature and annual humidity as important variables. When checking the response curves, it was found that both variables play positive roles in predicting the occurrence of *C. indicum*. That is, higher minimum temperature or annual humidity will result into higher probability of occurrence of the species.

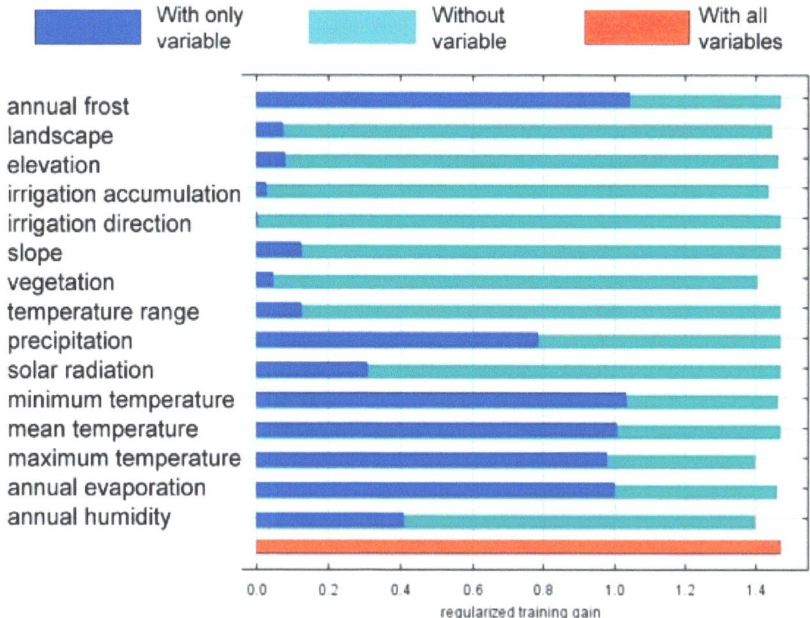

Figure 2: The jackknife of regularized training gain for *C. indicum*.

By using ecological niche modeling, this exemplary study showed that the most current suitable range of *C. indicum* could be found in the eastern and central part of China, with some marginal distribution in southern and southwestern part of China. Some of the most important correlated environmental variables, minimum temperature and annual humidity, were jointly identified by percent contribution, permutation importance and jackknife test based on the output of Maxent prediction.

REFERENCES

[1] N. Myers, R. Mittermeier, C. Mittermeier, G. da Fonseca, J. Kent, "Biodiversity hotspots for conservation priorities", *Nature*. 403, 853-858, 2000.

[2] V. Braunisch, R. Suchant, "A model for evaluating the "habitat potential" of a landscape for capercaillie Tetrao urogallus: a tool for conservation planning", *Wild. Biol*. 13, 21-33, 2007.

[3] F. Mestre, J. Ferreira, A. Mira, "Modelling the distribution of the European polecat Mustela putorius in a Mediterranean agricultural landscape", *Rev. Ecol. Terre Vie*. 62, 35-47, 2007.

[4] L. Traill, R. Bigalke, "A presence-only habitat suitability model for large grazing African ungulates and its utility for wildlife management", *Afr. J. Ecol*. 45, 347-354, 2006.

[5] A. Hirzel, B. Posse, P. Oggier, Y. Crettenands, C. Glenz, R. Arlettaz, "Ecological requirement of reintroduced species and the implications for release policy: the case of bearded vulture", *J. Appl. Ecol*. 41, 1103-1116, 2004.

[6] C. Soares, J. Brito, "Environmental correlates for species richness among amphibians and reptiles in a climate transition area", *Biodivers. Conserv*. 16, 1087-1102, 2007.

[7] J. Elith, S. Phillips, T. Hastie, M. Dudik, Y. Chee, C. Yates, "A statistical explanation of MaxEnt for ecologists", *Divers. Distrib.* 17, 43-57, 2011.

[8] S. Phillips, M. Dudik, "Modeling of species distributions with Maxent: new extensions and a comprehensive evaluation", *Ecography*. 31, 161-175, 2008.

[9] S. Phillips, R. Anderson, R. Schapire, "Maximum entropy modeling of species geographic distributio", in: C. Brodley, J. Cavazos, P. Chan, N. Jacob, A. Kak, R. Lippman, *et al.* (Eds.), Proc. 21st Int. Conf. Mach. Learn., ACM Press, New York, 2004: pp. 655-662.

[11] H. Nix, "A biogeographic analysis of Australian elapid snakes", in: R. Longmore (Ed.), Atlas Elapid Snakes Aust., Bureau of Flora and Fauna, Canberra, 1986: pp. 4-15.

[12] J. Busby, "BIOCLIM-a bioclimate analysis and prediction system", *Plant Prot.* Q. 6, 8-9, 1991.

[13] A. Hirzel, J. Hausser, D. Chessel, N. Perrin, "Ecological-Niche Factor Analysis: how to compute habitat-suitability maps without absence data?", *Ecology*. 83, 2027-2036, 2002.

[14] Y. Chen, "Influence of environment and space on haplotype composition structure of populations of Chrysanthemum indicum L. (Compositae) in China with a prediction of its suitable range", *J. Biodivers. Manag.* For. 2, 2, 2013.

[15] H. Fang, Q. Guo, H. Shen, Q. Shao, "Phylogeography of Chrysanthemum indicum L. (Copositae) in China based on trnL-F sequences", *Biochem. Syst. Ecol.* 38, 1204-1211, 2010.

[16] S. Dray, P. Legendre, P.R. Peres-Neto, "Spatial modelling: a comprehensive framework for principal coordinate analysis of neighbour matrices (PCNM)", *Ecol. Model.* 196, 483-493, 2006.

[17] P. Legendre, X. Mi, H. Ren, K. Ma, M. Yu, I.-F. Sun, *et al.,* "Partitioning beta diversity in a subtropical broad-leaved forest of China", *Ecology*. 90, 663-674, 2009.

[18] B. Hong, Y. Wang, H. Zhao, "Suitable distribution area of Eriosoma lanigerum (Hausmann) in China and related affecting factors", *Chin. J. Appl. Ecol.* 23, 1123-1127, 2012.

[19] B. Shipley, C.E.T. Paine, C. Baraloto, "Quantifying the importance of local niche-based and stochastic processes to tropical tree community assembly"., *Ecology*. 93, 760-9, 2012.

[20] G. Rodriguez-Castaneda, A. Hof, R. Janson, L. Harding, "Predicting the fate of biodiversity using species' distribution models: enhancing model comparability and repeatability", *PLoS ONE*. 7, e44402, 2010.

[21] A. Jimerez-Valverde, P. Acevedo, A. Barbosa, J. Lobo, R. Real, "Dsicrimination capability in species distribution models depends on the representativeness of the environmental domain", *Glob. Ecol. Biogeogr*. 22 508-516, 2013.

[22] Y. Chen, "Habitat suitability modeling of amphibian species in southern and central China: environmental correlates and potential richness mapping", *Sci. China Life Sci.* 56, 476-484, 2013.

[23] X. Kou, Q. Li, S. Liu, "Quantifying species' range shifts in relation to climate change: a case study of Abies spp. in China", *PLoS ONE*. 6, e23115. 2011

CHAPTER 13

Phylogenetic Relatedness Pattern and Climatic Correlates on the Distribution of Endemic Birds in China

Abstract: In this chapter, the mesoscale phylogenetic community structure of endemic birds of China is analyzed and the associated possible driving climatic variables are evaluated. The results show that, over the 568 $0.5^{\circ} \times 0.5^{\circ}$ grid cells covering the territory of China, overdispersion is the major phylogenetic community pattern for endemic birds, as indicated by prevailing negative values of standardized net relatedness index (*NRI*) and the nearest taxon index (*NTI*), and the small *NRI/NTI* ratio values over grid cells. The most correlated climatic variable with *NRI/NTI* ratio over studied grid cells is relative humidity as indicated by partial correlation test which controls the effect of spatial autocorrelation. Interspecific competition may explain the overdispersion patterns for endemic bird community structure in China.

Keywords: Avian endemism, clustering versus overdispersion, ecophylogenetics, environmental correlates, environmental filtering, evolutionary distinctiveness, evolutionary history, macroecology, phylogenetic community structure, phylogenetic filtering, phylogenetic history, spatial distribution.

INTRODUCTION

Phylogenetic community structure [1-4] has now become widely appreciated in community ecology by considering the impacts of phylogenetic history of species on structuring species composition patterns. Typically there are three types of phylogenetic community structure found in the contemporary literature: clustering, overdispersion and randomness [1, 2, 4].

Phylogenetic niche conservatism theory [5-9] predicts that closely related species tend to have similar functional traits. Such a prediction can be naturally extended to the distribution of species, which tend to distribute aggregately when they are closely related in the phylogenetic relationship. This aggregated distribution pattern is named as "phylogenetic clustering pattern" in phylogenetic community structure studies as mentioned above [1, 2, 10]. The opposite pattern is "phylogenetic overdispersion pattern", where closely related species tend to distribute in different areas without too many range overlaps because of competition exclusion. Both patterns can be observed for various taxa in previous studies [11-14]. Finally, Phylogenetic randomness is a kind of pattern without

Youhua Chen

phylogenetic signals. Thus, phylogenetic history plays no roles on structuring species distribution and associated community composition patterns.

Phylogenetic relatedness pattern of specific species assemblages at microcosm, local or broad-spatial scales are now gaining attention and widely studied in many taxonomic groups [15-20]. Currently, it is still unclear whether a subset of a taxonomic group (endemic species) can present remarkable phylogenetic relatedness pattern in comparison to the case where all species in that taxonomic group are included for phylogenetic community structure analysis. Because many endemic species are speciated to occur in relatively young evolutionary time in comparison to other common taxa (of course, some old species can become endemic due to the local extinction of ranges outside the studied area), we would expect that phylogenetic clustering pattern is predominant for endemic species because of their relatively limited dispersal abilities.

Moreover, contemporary distribution of endemic species is driven by multiple and complex variables. Environmental gradients are recognized as one of main forces driving the distribution of species [21-23]. In recent studies, it is found that some environmental variables are strongly correlated with phylogenetic relatedness patterns of species assemblages over various spatial scales [12, 13, 15]. So, what will be the relationship between phylogenetic community structure and climatic variability for endemic species assemblages? In the present study, by investigating the distribution of endemic birds in China, we will test the above-mentioned predictions relevant to phylogenetic relatedness patterns of endemic taxa.

MATERIALS AND METHODS

Construction of Phylogenetic Tree and Distribution of Endemic Birds in China

The list of the endemic birds of mainland China was gathered from previous studies [24-26] and World Bird Database (http://avibase.bsc-eoc.org/). Table **1** listed out the species for the present study. A list of the studied species' names is presented in Table **1**. Distributional records of each endemic bird are retrieved from previous studies, resulting into 568 quadrats each of which has a resolution of 0.5° latitude $\times$ 0.5° longitude. Measurement of phylogenetic relatedness and climatic determinants are both done on these quadrats.

Measurement of Phylogenetic Relatedness

I calculate two indices, the standardized net relatedness index (*NRI*) and the nearest taxon index (*NTI*) for the distribution of endemic bird species across different grid cells of China to measure their phylogenetic relatedness. These two indices [1, 3] have been widely applied to phylogenetic community structure analysis:

$$NRI = -\frac{MPD - rndMPD}{sd(rndMPD)}, \quad (1)$$

$$NTI = -\frac{MNTD - rndMNTD}{sd(rndMNTD)}. \quad (2)$$

where MPD represents the mean pairwise phylogenetic distance in which it finds the average distance to all other taxa in the sample for each taxon and $MNTD$ is to calculate the nearest phylogenetic neighbor in the sample for each taxon [3]. The *rndMPD* and *rndMNTD* represent the mean *MPD* and mean *MNTD* from randomly generated assemblages. Negative values of both metrics indicate overdispersion, while positive values of both metrics indicate clustering [27]. These two standardized indices should follow the unit normal distribution with mean=0 and variance=1. As such, if any of the indices is larger than 1.96 (or less than -1.96), the clustering (or overdispersion) pattern identified is regarded to be statistically significant.

However, *NRI* is found to be biased when detecting overdispersion pattern [28, 29], thus the ratio *NRI/NTI* is used to quantify overdispersion or clustering patterns by comparing the observed ratio to the random ones. If no more than 5% of randomly simulated *NRI/NTI* values larger than the observed one (that is, *Prob(simulated<observed)>0.95*), phylogenetic clustering is recognized for the community. In contrast, if no more than 5% of random *NRI/NTI* values less than the observed one *(Prob(simulated>observed)>0.95)*, phylogenetic overdispersion is suggested accordingly [30].

For all the quadrats being studied, the assembled *NTI* and *NRI* values are tested using one-sample *t*-test to test whether they are significantly larger or smaller than zero since it is normally distributed. If the observed values are significantly larger (or smaller) than zero, an overall phylogenetic clustering (or overdispersion) trend across communities are suggested [30].

Climatic Correlates of Phylogenetic Community Structure

The following climatic variables are considered: minimum annual temperature, maximum annual temperature, mean annual temperature, evaporation, radiation, precipitation and humidity. For removing the impacts of spatial autocorrelation, a partial correlation test is utilized to test the association between the ratio *NRI/NTI* and climatic variables.

RESULTS

Endemic Birds of China Showed Phylogenetic Overdispersion Pattern

For *NRI/NTI* ratio, it is observed that over 568 grid cells, 365 ones have smaller values than the mean ratio while only 203 ones have larger values than the mean ratio. For NTI and NRI indices, 342 and 349 quadrats have negative values respectively. Correspondingly, 226 and 219 ones have positive values.

Moreover, one sample *t*-test showed that, the mean NRI value (= -0.1) is significantly less than zero ($p<0.05$), while the mean NTI value (= -0.04) is not significantly different from zero ($p=0.28$). As such, it can be concluded that weak phylogenetic overdispersion is the prevailing phylogenetic community structure for endemic birds of China.

Important Environmental Variables that May Influence the Phylogenetic Community Structure

Based on the results of partial correlation test (Table **1**) by controlling the influence of spatial autocorrelation, it could be found that humidity is the most important variable linked to phylogenetic relatedness of species assemblages (partial correlation coefficient $r=0.382$, $P<0.05$), followed by annual evaporation (partial $r=-0.323$, $P<0.05$). The role of temperature variables is also important with partial r around 0.3 except for the annual maximum temperature, which has a relatively smaller $r=-0.17$ but is still significant ($P<0.05$). Fig. **1** showed the scatter point patterns between these climatic variables and the *NRI/NTI* ratio.

DISCUSSION

The prediction that endemic species might possess phylogenetic clustering pattern for their distribution relies on the assumptions of their limited dispersal abilities and limited time to colonize new habitats, as mentioned above. However, these assumptions are not enough to drive endemic species to be phylogenetically clustered in terms of distribution because if these species are originated from different areas which are not overlapped and isolated from each other.

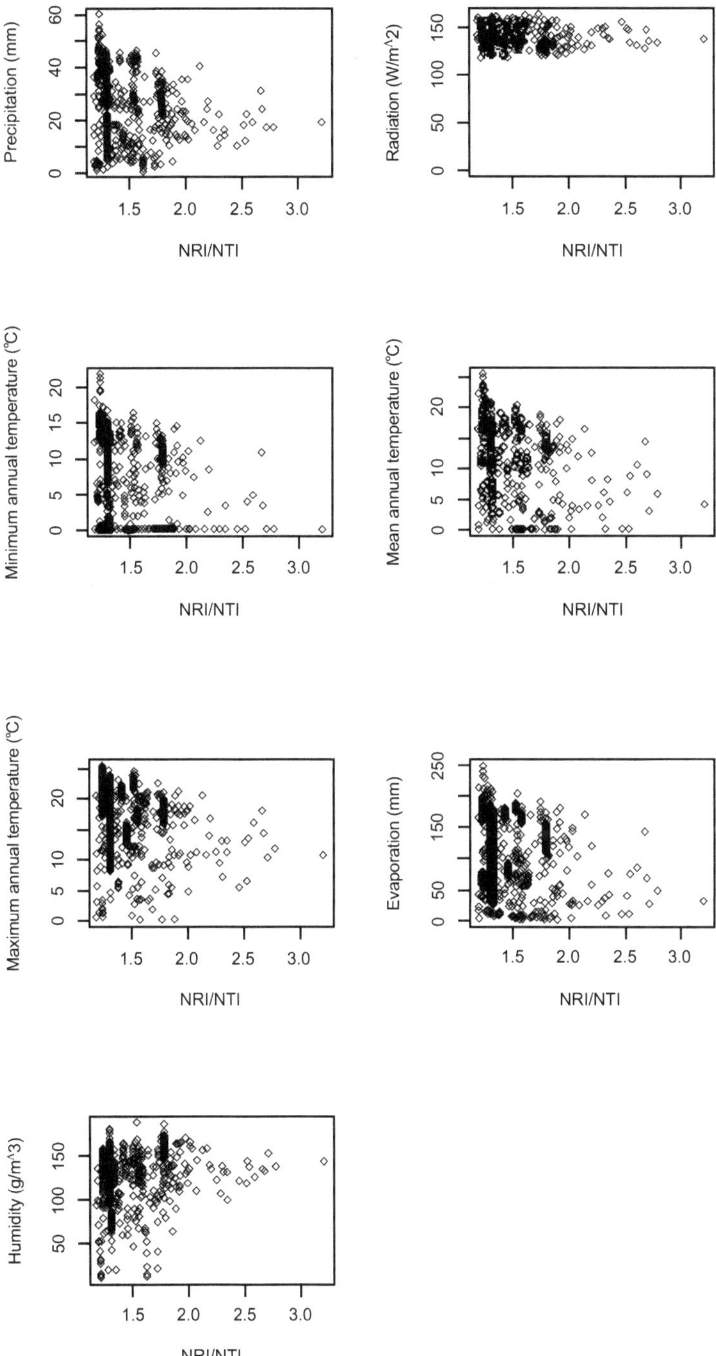

Figure 1: Plots between *NRI/NTI* ratio and climatic variables for the distribution of endemic birds of China.

Table 1: Partial correlation between *NRI/NTI* ratio and climatic variables by controlling the effect of spatial autocorrelation

	NRI/NTI	*P*
Precipitation	-0.169	<0.05
Radiation	-0.133	<0.05
Minimum annual temperature	-0.308	<0.05
Mean annual temperature	-0.305	<0.05
Maximum annual temperature	-0.17	<0.05
Evaporation	-0.323	<0.05
Humidity	0.382	<0.05

Based on the prediction of "out-of-tropics" hypothesis, tropic areas should harbour more endemic species because of high speciation rates there. As such, it is expected that phylogenetic clustering patterns in tropical areas should be more prevalent in comparison to other areas like temperate zones. Temperate zones have less signature of clustering patterns because most of species there, are originated from tropical zones, resulting into ecological communities with species from different clades. Consequently, the likelihood of phylogenetic overdispersion pattern of species communities in temperate zones is much higher than that in tropical zones. Therefore, it is natural to predict that there is a positive relationship between relatedness patterns and temperature variables. However, as seen in Table **1** and Fig. **1**, there is a positive correlation between temperature-related variables and the *NRI/NTI* ratio. Thus, being contradictory to the prediction above, warmer areas will have increasing phylogenetic dispersion trend. This interesting pattern might be due to the fact that for endemic taxa, they might have strong interspecific competition from each other, resulting into the overdispersion pattern in warm areas.

Overall, our study showed that phylogenetic random pattern is dominant for the distribution of endemic bird species over the territory of China. Humidity is found to be the most important climatic variable correlated with phylogenetic community patterns of these species.

REFERENCES

[1] C. Webb, D. Ackerly, M. Mcpeek, M. Donoghue, "Phylogenies and community ecology", *Annu. Rev. Ecol. Syst.* 33, 475-505, 2002.

[2] N. Kraft, W. Cornwell, C. Webb, D. Ackerly, "Trait evolution, community assembly, and the phylogenetic structure of ecological communities", *Am. Nat.* 170, 271-283, 2007.

[3] C. Webb, D. Ackerly, S. Kembel, "Phylocom: software for the analysis of phylogenetic community strucutre and trait evolution", *Bioinformatics*. 15, 2098-2100, 2008.

[4] J. Cavender-Bares, D. Ackerly, D. Baum, F. Bazzaz, "Phylogenetic overdispersion in Floridian oak communities", *Am. Nat.* 163, 823-843, 2004.

[5] J. Losos, "Phylogenetic niche conservatism, phylogenetic signal and the relationship between phylogenetic relatedness and ecological similarity among species", *Ecol. Lett.* 11, 995-1003, 2008.

[6] H. Liu, E. Edwards, R. Freckleton, C. Osborne, "Phylogenetic niche conservatism in C4 grasses", *Oecologia*. 170, 835-845, 2012.

[7] J. Wiens, "Speciation and ecology revisited: phylogenetic niche conservatism and the origin of species", *Evolution*. 58, 193-197, 2004.

[8] M. Crisp, L. Cook, "Phylogenetic niche conservatism: what are the underlying evolutionary and ecological causes?", *New Phytol.* 196, 681-694, 2012.

[9] C. Hof, C. Rahbek, M. Araujo, "Phylogenetic signals in the climatic niches of the world's amphibians", *Ecography*. 33, 242-250, 2010.

[10] C. Webb, C. Cannon, S. Davies, "Ecological organization, biogeography and the phylogenetic structure of tropical forest tree communities, in: W. Carson, S. Schnitzer (Eds.)", *Trop. For. Community Ecol.*, Blackwell, Oxford, 2008.

[11] N. Cooper, J. Rodriguez, A. Purvis, "A common tendency for phylogenetic overdispersion in mammalian assemblages", PRSB. 275 (2008) 2031-2037.

[12] X. Li, X. Zhu, Y. Niu, H. Sun, "Phylogenetic clustering and overdispersion for alpine plants along elevational gradient in the Hengduan Mountains Region, Southwest China", *J. Syst. Evol.* In press 2013.

[13] J. Wang, J. Soininen, J. He, J. Shen, "Phylogenetic clustering increases with elevation for microbes", *Environ. Microbiol. Rep.* 4, 217-226, 2012.

[14] M. Horner-Devine, B. Bohannan, "Phylogenetic clustering and overdispersion in bacterial communities", *Ecology*. 87, S100-S108, 2006.

[15] H. Qian, Y. Zhang, J. Zhang, X. Wang, "Latitudinal gradients in phylogenetic relatedness of angiosperm trees in North America", *Glob. Ecol. Biogeogr.* In press 2013.

[16] J. Tan, Z. Pu, W. Ryberg, L. Jiang, "Species phylogenetic relatedness, priority effects, and ecosystem functioning", *Ecology*. 93, 1164-1172, 2012.

[17] K. Peay, M. Belisle, T. Fukami, "Phylogenetic relatedness predicts priority effects in nectar yeast communities", *Proc. R. Soc. B Biol. Sci.* 279, 749-758, 2012.

[18] N. Swenson, B. Enquist, J. Thompson, J. Zimmerman, "The influences of spatial and size scale on phylogenetic relatedness in tropical forest communities", *Ecology*. 88, 1770-1780, 2007.

[19] M. Cadotte, M. Hamilton, B. Murray, "Phylogenetic relatedness and plant invader success across two spatial scales", *Divers. Distrib.* 15, 481-488, 2009.

[20] Y. Chen, "Distributional patterns of alien plants in China: the relative importance of phylogenetic history and functional attributes", *ISRN Ecol.* 2013 (2013) 527052.

[21] A. Jimenez-Valverde, N. Barve, A. Lira-Noriega, S. Maher, Y. Nakazawa, M. Papes, *et al.*, "Dominant climate influences on North American bird distributions", *Glob. Ecol. Biogeogr.* 20, 114-118, 2011.

[22] J. Zhang, W. Kissling, F. He, "Local forest structure, climate and human disturbance determine regional distribution of boreal bird species richness in Alberta, Canada", *J. Biogeogr.* (2012) doi:10.1111/jbi.12063.

[23] M. Munguia, C. Rahbek, T. Rangel, J. Diniz-Filho, M. Araujo, "Equilibrium of global amphibian species distributions with climate", Plos One. 7 (2012) e34420.

[24] F. Lei, T. Lu, China endemic birds, Science Press, Beijing, 2006.

[25] F. Lei, J. Lu, Y. Liu, Y. Qu, Z. Yin, "Endemic bird species to China and their distribution", *Curr. Zool.* 48, 599-610, 2002.

[26] F. Lei, G. Wei, H. Zhao, Z. Yin, J. Lu, "China subregional avian endemism and biodiversity conservation", *Biodivers. Conserv.* 16, 1119-1130, 2007.

[27] N. Pei, J. Lian, D. Erickson, N. Swenson, W. Kress, W. Ye, *et al.*, "Exploring tree-habitat associations in a Chinese subtropical forest plot using a molecular phylogeny generated from DNA barcord loci", *Plos One*. 6 (2011) e21273.

[28] S. Kembel, S. Hubbell, "The phylogenetic structure of a neotropical forest tree community", *Ecology*. 87, S86-S99, 2006.

[29] N. Swenson, B. Enquist, J. Pither, J. Thompson, J. Zimmerman, "The problem and promise of scale dependency in community phylogenetics", *Ecology*. 87 2418-2424, 2006.

[30] N. Cooper, J. Rodriguez, A. Purvis, "A common tendency for phylogenetic overdispersion in mammalian assemblages", *PRSB*. 275, 2031-2037, 2008.

CHAPTER 14

Quantifying the Relative Contribution of Climatic Variability and Dispersal Limitation on the Distribution of Endemic Birds of Mainland China Using Spatial Point Pattern Analysis

Abstract: In this chapter, the relative importance of climatic variability and dispersal limitation on the distribution of endemic birds in mainland territorial region of China is analyzed using inhomogeneous Poisson process models. Model comparison is performed through Akaike Information Criteria (AIC). The results showed that, over the 42 species studied, climatic variability was more important than dispersal limitation for explaining the spatial distributional patterns of 6 species. In contrast, the spatial distribution of 10 species was better explained by the dispersal limitation mechanism. Both climatic variability and dispersal limitation play equal roles on structuring the spatial distributional record patterns of the remaining 26 endemic bird species. In conclusion, different species responses climatic fluctuation in various ways and their migration over neighboring regions may result into the diversified influence of dispersal limitation and climatic variability. Dispersal limitation seems more important than climatic variability on structuring the distribution of endemic taxa, at least for the endemic birds of mainland China.

Keywords: Akaike Information Criteria, avian endemism, climatic niche, dispersal limitation, distributional patterns, endemic species, homogeneous processes, model comparison, niche versus neutrality, spatial ecology, spatial statistics.

INTRODUCTION

Contemporary distribution of species is driven by multiple factors. Two important factors are dispersal limitation and environmental filtering [1-9]. Dispersal limitation is relevant to the species' migration capability while habitat heterogeneity implies the influence of environment.

There are some methods to quantify relative importance of habitat filtering and dispersal limitation for ecological data. For example, variation partitioning based on species compositional matrix [8, 10-13], variation partitioning based on the raw distribution of species using spatial point analysis [1, 2, 14-17]. Different methods have different advantages. For example, spatial point analysis [16] does not require spatial sampling quadrats, thus becoming invariant over the change of spatial scales. Also, another merit of this method is that it can reveal the relative

Youhua Chen

importance of different ecological mechanisms on each single species. In contrast, species compositional data-based regression analysis [8, 10] should have sampling quadrats for building the species-site matrix for subsequent analyses. Such a method can not identify the relative influence of different mechanisms on structuring the distribution of each single species (but it can be applied for haplotype distributional data, see [18]). However, spatial point analysis is more complex than regression-based variation partitioning method.

Endemic taxa are always focused in conservation and biogeographic researches [19-23]. In the present study, I apply spatial point pattern analysis for revealing the relative influence of dispersal limitation and climatic variability on structuring the distribution of endemic birds in mainland China given the fact that no previous studies have done this research yet. The reason for using spatial point pattern analysis instead of regression-based variation partitioning method is that I want to identify the relative influence of both ecological processes on the distribution of each single endemic bird species over the territory of mainland China.

MATERIALS AND METHODS

Distribution of Endemic Birds of China

An endemic bird species list of mainland China (territorial region, but exclude islands like Hainan and Taiwan) was gathered from previous studies [24-26] and World Bird Database (http://avibase.bsc-eoc.org/). Distributional records of these species are gathered from the Chinese Species Information System (Chinese website) (http://csis.baohudi.org/csis_search/index.php). Some species are excluded for the present study because their distribution has no more than 1 geographic record, resulting 42 species for the present study. The checklist of these 42 species is presented in Table **1**.

Variables Indicating Climatic Fluctuation

Based on some previous studies [27-29], the following variables are selected as the indicators of climatic variability: the annual minimum temperature (MinTemp), maximum temperature (MaxTemp), mean temperature (MeanTemp) and annual precipitation (Prec).

Variables Representing Dispersal Limitation

Similar to previous studies [8, 12, 30], I used latitude (Lat) and longitude (Long) as the indicators of dispersal limitation.

Spatial Point Analysis

I used spatial point analysis to quantify the relative importance of climatic variablity and dispersal limitation. The inhomogeneous Poisson process (IPP) is constructed using the above variable groups as the explanatory variables while the intensity of spatial distributional points of endemic species is used as a response variable.

For the IPP model accounting for dispersal limitation, the model is: for $\lambda(\mu)$, the mean number of events (spatial points) per unit area at a location μ, it is estimated to relate to latitude and longitude as $\lambda(\mu) = \exp(\theta_0 + \theta_1 \times \mathrm{Lat} + \theta_2 \times \mathrm{Long})$. I called this as Model 0 for subsequent analyses.

For the IPP model accounting for climatic variability, the mean number of spatial points per unit area at location μ is estimated as $\lambda(\mu) = \exp(\theta_0 + \theta_1 \times \mathrm{MinTemp}$ $+\theta_2 \times \mathrm{MaxTemp} + \theta_2 \times \mathrm{MeanTemp} + \theta_2 \times \mathrm{Prec})$. This is called as Model 1 for comparison.

For my study, for each species, both Models 0 and 1 are fitted onto its spatial distributional patterns. Then, I utilize Akaike Information Criteria (AIC) [31] to select the better model from the two comparable models. Lower AIC the model has, the better it is. If the AIC difference $\Delta\,\mathrm{AIC}$ between the two models are more than 10 [32, 33], the one with lower AIC is believed to be significantly favoured.

RESULTS

Through the model comparison using AIC criteria (Table **1**), it was identified that over 42 endemic bird species for the study, 6 species were better explained by climatic variablity, including *Arborophila gingica, Garrulax elliotii, Garrulax maximus, Lophophorus lhuyssi, Sitta yunnanensis,* and *Tragopan caboti*. In contrast, 10 species were better quantified by dispersal limitations (Table **1**), which were *Aegithalos fuliginosus, Alectoris magna, Chrysolophus pictus, Crossoptilon auritum, Latoucheornis siemsseni, Liocichla omeiensis, Phoenicurus alaschanicus, Podoces biddulphi, Rhopophilus pekinensis, and Syrmaticus reevesii*. The other 26 species show preferences on neither spatial nor climatic variables, implying that both dispersal limitation and climatic variability have of equal importance on structuring the distribution of these species.

Table 1: AIC comparison between inhomogeneous Poisson process models accounting for dispersal limitation (Model 0) and climatic variability (Model 1)

Species	AIC (model 0)	AIC (model 1)	Δ AIC
Aegithalos fuliginosus	254.641	224.146	30.495
Alcippe striaticollis	167.221	172.762	-5.541
Alcippe variegaticeps	91.74	87.81	3.93
Alectoris magna	107.721	88.597	19.124
Arborophila gingica	133.498	153.116	-19.618
Arborophila rufipectus	77.959	77.936	0.023
Babax koslowi	65.159	67.732	-2.573
Bonasa sewerzowi	215.284	220.668	-5.384
Carpodacus eos	291.951	299.078	-7.127
Carpodacus roborowskii	33.848	35.315	-1.467
Chrysolophus pictus	847.356	515.123	332.233
Crossoptilon auritum	454.56	443.963	10.597
Crossoptilon mantchuricum	319.648	328.159	-8.511
Emberiza koslowi	44.898	42.743	2.155
Garrulax bieti	39.893	46.87	-6.977
Garrulax davidi	109.344	115.676	-6.332
Garrulax elliotii	477.993	504.846	-26.853
Garrulax lunulatus	63.581	68.065	-4.484
Garrulax maximus	229.417	241.057	-11.64
Garrulax sukatschewi	87.422	88.737	-1.315
Latoucheornis siemsseni	147.1	105.108	41.992
Leptopoecile elegans	162.99	159.728	3.262
Liocichla omeiensis	104.279	76.834	27.445
Lophophorus lhuysii	267.893	284.425	-16.532
Oriolus mellianus	68.86	71.933	-3.073
Paradoxornis conspicillatus	33.361	36.443	-3.082
Paradoxornis paradoxus	42.689	36.214	6.475
Paradoxornis przewalskii	46.157	46.169	-0.012
Paradoxornis zappeyi	87.338	88.886	-1.548
Parus davidi	146.984	147.994	-1.01
Parus superciliosus	219.711	210.84	8.871
Perisoreus internigrans	120.707	121.238	-0.531
Phoenicurus alaschanicus	159.545	141.011	18.534

Table 1: contd...

Phylloscopus kansuensis	34.66	37.624	-2.964
Podoces biddulphi	143.77	116.167	27.603
Rhopophilus pekinensis	188.318	175.611	12.707
Sitta yunnanensis	153.859	190.034	-36.175
Strix davidi	93.077	96.905	-3.828
Syrmaticus reevesii	547.776	338.903	208.873
Tetraophasis obscurus	178.863	187.138	-8.275
Tragopan caboti	173.522	191.916	-18.394
Urocynchramus pylzowi	166.245	156.518	9.727

DISCUSSION

Usually, in community ecology, the relative influence of dispersal limitation and environmental heterogeneity is modelled using regression-based methods, like redundancy analysis or canonical correspondence analysis [8, 9, 34, 35]. Spatial point pattern analysis offers a new way [14] to explore the influence of these ecological mechanisms on each single species from the whole ecological community and have been successfully applied in some previous studies [1, 2]. The intensity of spatial points (*i.e.*, distributional records of each species here) $\lambda(\mu)$ is the response variable when performing point process modelling.

As showed in Table **1**, the relative contribution of climatic variability and dispersal limitation on the distribution of each endemic bird species in mainland China varied greatly. Over the 42 endemic species being studied, the distribution of six species is better quantified by climatic variability while the distribution of ten species is better predicted by dispersal limitation. As such, it seems that dispersal limitation is a more important process for structuring the distribution of endemic taxa. This really makes senses because the definition of endemic taxa is that they usually have very narrow spatial distributional ranges. As such, limited dispersal capability for these taxa is highly desired.

Currently, there are not many empirical studies [1, 2] applying spatial point pattern analysis to understand the influences of different ecological mechanisms on each single species and the whole community. More theoretical development and empirical applications are required on this regard so as to better expand and improve the application of spatial statistics on quantifying the relative importance of ecological processes structuring species' distributional patterns.

REFERENCES

[1] G. Shen, M. Yu, X. Hu, X. Mi, H. Ren, I. Sun, *et al.*, "Species-area relationships explained by the joint effects of dispersal limitation and habitat heterogeneity", *Ecology*. 90, 3033-3041, 2009.

[2] G. Shen, F. He, R. Waagepetersen, I. Sun, Z. Hao, Z. Chen, *et al.*, "Quantifying effects of habitat heterogeneity and other clustering processes on spatial distributions of tree species", *Ecology*. (2013) In press.

[3] J. Franklin, G. Keppel, E. Webb, J. Seamon, S. Rey, D. Steadman, *et al.*, "Dispersal limitation, speciation, environmental filtering and niche differentiation influence forest tree communities in West Polynesia", *J. Biogeogr.* 40, 988-999, 2013.

[4] M. Réjou-Méchain, O.J. Hardy, "Properties of similarity indices under niche-based and dispersal-based processes in communities", *Am. Nat.* 177 (2011) 589-604. doi:10.1086/659627.

[5] A. Algar, D. Mahler, R. Glor, L. Losos, "Niche incumbency, dispersal limitation and climate shape geographical distributions in a species-rich island adaptive radiation", *Glob. Ecol. Biogeogr.* 22, 391-402, 2013.

[6] A. Freestone, B. Inouye, "Dispersal limitation and environmental heterogeneity shape scale-dependent diversity patterns in plant communities", *Ecology*. 87, 2425-2432, 2006.

[7] B. Shipley, C.E.T. Paine, C. Baraloto, "Quantifying the importance of local niche-based and stochastic processes to tropical tree community assembly", *Ecology*. 93, 760-9, 2012.

[8] D. Borcard, P. Legendre, P. Drapeau, "Partialling out the Spatial Component of Ecological Variation", *Ecology*. 73, 1045, 1992.

[9] P. Peres-Neto, P. Legendre, S. Dray, D. Borcard, "Variation partitioning of species data matrices: estimation and comparison of fractions", *Ecology*. 87, 2614-2625, 2006.

[10] P. Legendre, L. Legendre, Numerical ecology, Elsevier Science BV, Amsterdam, 1998.

[11] A. Meot, P. Legendre, D. Borcard, "Partialling out the spatial component of ecological variation: questions and propositions in the linear modelling framework", *Environ. Ecol. Stat.* 5, 1-27, 1998.

[12] P. Legendre, D. Borcard, P. Peres-Neto, "Analyzing beta diversity: Partitioning the spatial variation of community composition data", *Ecol. Monogr.* 75, 435-450, 2005.

[13] P. Legendre, "Studying beta diversity: ecological variation partitioning by multiple regression and canonical analysis", *J. Plant Ecol.* 1, 3-8, 2007.

[14] A. Jalilian, Y. Guan, R. Waagepetersen, "Decomposition of variance for spatial Cox processes", *Scand. J. Stat.* 40, 119-137, 2013.

[15] R. Waagepetersen, Y. Guan, "Two-step estimation for inhomogeneous spatial point processes", *J. R. Stat. Soc. Ser. B.* 71, 685-702, 2009.

[16] P. Diggle, Statistical analysis of spatial point patterns, Academic Press, London, UK, 2003.

[17] S.M. Eckel, Statistical Analysis of Spatial Point Patterns: Applications to Economical, Biomedical and Ecological Data, Universität Ulm 2008, 2008.

[18] Y. Chen, "Influence of environment and space on haplotype composition structure of populations of Chrysanthemum indicum L. (Compositae) in China with a prediction of its suitable range", *J. Biodivers. Manag. For.* 2, 2, 2013.

[19] Y. Chen, "Modeling extinction risk of endemic birds of mainland China", *Int. J. Evol. Biol.* 2013 (2013) 639635.

[20] Y. Chen, "A phylogenetic subclade analysis of range sizes of endemic woody see plant species of China: trait conservatism, diversification rates and evolutionary models", *J. Syst. Evol.* 51, 590-600, 2013.

[21] J. Malcolm, C. Liu, R. Neilson, L. Hansen, L. Hannah, "Global warming and extinctions of endemic species from biodiversity hotspots", *Conserv. Biol.* 20 538-548, 2006.

[22] J. Huang, J. Chen, J. Ying, K. Ma, "Features and distribution patterns of Chinese endemic seed plant species", *J. Syst. Evol.* 49, 81-94, 2011.

[23] A. Grill, R. Crnjar, P. Casula, S. Menken, "Applying the IUCN threat categories to island endemics: Sardinian butterflies (Italy)", *J. Nat. Conserv.* 10, 51-60, 2002.

[24] F. Lei, T. Lu, China endemic birds, Science Press, Beijing, 2006.

[25] F. Lei, J. Lu, Y. Liu, Y. Qu, Z. Yin, "Endemic bird species to China and their distribution", *Curr. Zool.* 48, 599-610, 2002.

[26] F. Lei, G. Wei, H. Zhao, Z. Yin, J. Lu, "China subregional avian endemism and biodiversity conservation", *Biodivers. Conserv.* 16, 1119-1130, 2007.

[27] H. Qian, W. Kissling, X. Wang, P. Andrews, "Effects of woody plant species richness on mammal species richness in southern Africa", *J. Biogeogr.* 36, 1685-1697, 2009.

[28] H. Qian, W. Kissling, "Spatial scale and cross-taxon congruence of terrestrial vertebrate and vascular plant species richness in China", *Ecology.* 91, 1172-1183, 2010.

[29] J. Zhang, W. Kissling, F. He, "Local forest structure, climate and human disturbance determine regional distribution of boreal bird species richness in Alberta, Canada", *J. Biogeogr.* (2012) doi:10.1111/jbi.12063.

[30] B. Gilbert, J.R. Bennett, "Partitioning variation in ecological communities: do the numbers add up?", *J. Appl. Ecol. 47* (2010) 1071-1082. doi:10.1111/j.1365-2664.2010.01861.x.

[31] H. Akaike, Information theory as an extension of the maximum likelihood principle, in: B. Petrov, F. Csaki (Eds.), Second Int. Symp. Inf. Theory, Akademiai Kiado, Budapest, 1974: pp. 276-281.

[32] Y. Chen, "An autoregressive model for global vertebrate richness rankings: long-distance dispersers could have stronger spatial structures", *Zool. Stud.* 52, 57, 2013.

[33] C. Vieira, Blmires, J. Diniz-Fiho, L. Bini, T. Rangel, "Autoregressive modelling of species richness in the Brazilian Cerrado", *Braz. J. Biol.* 68, 233-240, 2008.

[34] P. Legendre, X. Mi, H. Ren, K. Ma, M. Yu, I.-F. Sun, *et al.*, "Partitioning beta diversity in a subtropical broad-leaved forest of China", *Ecology.* 90, 663-674, 2009.

[35] Y. Chen, "A multiscale variation partitioning procedure for assessing the influence of dispersal limitation on species rarity and distribution aggregation in the 50-Ha tree plots of Barro Colorado Island, Panama", *J. Ecosyst. Ecography.* 3, 134, 2013.

Chapter 15: (A Phylogenetic Ancestral Endemism Index (PAE Incorporating the Information of Ancestral Ranges for Setting Conservation Priority of Species and Areas

CHAPTER 15

A Phylogenetic Ancestral Endemism Index (PAE) Incorporating the Information of Ancestral Ranges for Setting Conservation Priority of Species and Areas

Abstract: In the present chapter, a phylogenetic ancestral endemism index (PAE) was proposed to effectively incorporate the information derived from ancestral ranges of species. As a comparison, other previously proposed phylogenetic indices, including evolutionary distinctiveness (ED), taxonomic distinctiveness (TD), phylogenetic endemism (PE) and node-based I and W indices were all implemented. Distribution of eleven Psychotria endemic plants in four island groups of Hawaii was used as an example to evaluate and compare differences between PAE and other phylogenetic diversity indices when setting conservation priorities of species. My results showed that PAE is closely related to ED index, but distinct to other indices. PAE might be the only phylogenetic diversity index incorporating the information of ancestral states to evaluate conservation importance of species currently.

Keywords: Ancestral ranges, ancestral state reconstruction, conservation priority, evolutionary heritage, Hawaii islands, historical biogeography, phylogenetic diversity, phylogenetic endemism, phylogeography, plant endemism, species evolution.

INTRODUCTION

Phylogenetic diversity, for capturing the genetic variation and evolutionary history of species, has gaining much attention in recent years in conservation biogeography [1-5]. In comparison to other conventional biodiversity indices including species diversity, functional diversity [6] or environmental diversity [7-9], phylogenetic diversity could offer a unique way to evaluate the influences of historical evolution on structuring contemporary biodiversity patterns [4, 5, 10]. Although functional diversity may also be linked to the evolution of species [11], but the information typically is restricted to the ecological adaptation, but long-time speciation/extinction history could be not taken into account in functional diversity.

Simply, the phylogenetic diversity of a community is calculated by only considering the subtree for linking all the external species. Differently, phylogenetic diversity of a species could have a variety of definitions, if I only

Youhua Chen

consider the unique evolutionary history of a species, I could simply take the directed branch leading to that species (which is called as pendant branch) [12]. In many situations, phylogenetic distinctiveness [13-15] is widely used, which incorporating the unique history of the ancestors of species as parts of the total phylogenetic diversity accounted for that species.

At another perspective, endemic or distribution-narrow species is also very crucial in determining and identifying the extinction risk of species [16-18]. Typically geographically range-narrow species will have a narrow spectrum of tolerance on environmental change. Thus, they might face a greater likelihood of extinction threats in the near future when global change and anthropogenic activities increase the disturbance of these species. Thus, in recent studies, researchers tend to combine both branch length and distributional range information together to obtain the phylogenetic endemism index (PE) [19, 20], in which the weighting is the total union of ranges of species [19].

However, I will show in below that the index might have a problem when using the joint distributional range size of descended external species from a common ancestor (node) to weight the internal branch length directly led by that node. This is because in the history of this ancestor lineage, its actual distributional range is not the joint distributional ranges from all its descendents. This, from the perspective of historical biogeography, it is not appropriate to estimate the endemism or rarity of any internal ancestral lineages with the unreal range simply merged from external descendants.

As such, in the present chapter, I developed a new index, called phylogenetic ancestral endemism index (PAE), for the purpose to overcome the above question confronted by PE index. The new index requires the estimation of ancestral ranges for the internal ancestral lineages, which in turn could effectively contain and truly reflect the biogeographic history of species range evolution.

MATERIALS AND METHODS

Phylogenetic Endemism Index

PE is defined by considering two attributes: branch lengths and distributional ranges of external species [19, 20]. The calculation of phylogenetic endemism is illustrated as the following equation. For a given external (or tip) species,

$$PE_s = \sum_{i \in N(s)} l_i / R_i \qquad (1)$$

where $N(s)$ is the node set for species which formed a unique path from the root to the tip species s. l_i is the branch length led by node i, R_i is the union of distributional ranges of species which were the descendants from the node i. As such, PE is simply the summation of weighted branch lengths which linked the root to the tip species. The weighting is the range size of species.

A Problem in Phylogenetic Endemism Index

The PD index derived from equation (1) is quite straightforward, for any nodes, the weight is to count the distributional range of descended external species that share the node as the most recent common ancestor.

As shown in Fig. **1**, the internal node i (or ancestor species i) has the branch l_i and have three descended external species. Based on the calculation of PE, the weighted value for this internal branch should be l_i / R_i. However, R_i is not the true range of ancestor species i, thus didn't reflect the range size history of the ancestor. In this case, the range history of ancestor i has been over-estimated (under-estimation situation is also possible when the ancestral range is larger than the current joint range, but should be quite rare).

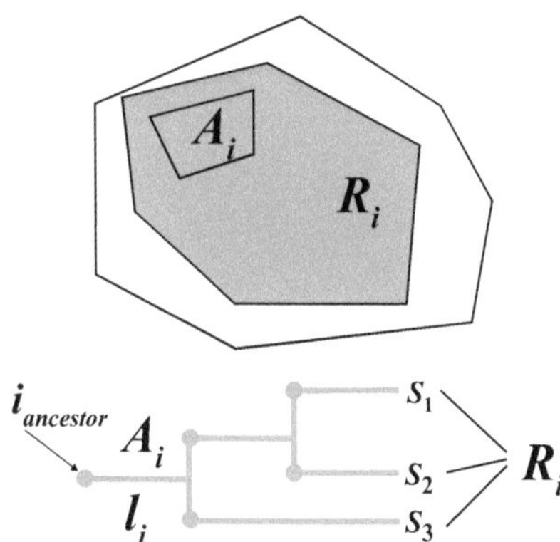

Figure 1: A problem of using contemporary range sizes of descendant lineages as the weight of internal nodes in phylogenetic endemism index. As seen, the ancestor species i (started from the node depicted by an arrow) has the ancestral distributional range A_i, while the three descended external species s_1, s_2 and s_3 have the joint distributional range of R_i. The internal branch length is l_i. As such, based on the calculation of PE (equation (1)), the weighted value for this internal branch should be l_i / R_i. However, R_i is not the true range of ancestor species i, thus didn't reflect the range size history of the ancestor. In this case, the range history of ancestor i has been over-estimated (under-estimation situation is also possible when the ancestral range is larger than the current joint range). Therefore, it is suggested to use the ancestral range A_i to indicate the true evolutionary history of species in that range. In view of this, I quantify phylogenetic ancestral endemism as l_i / A_i for the focusing branch.

A New Index Incorporating Ancestral Ranges of Extinct Lineages in Internal Nodes

When discussing the above problem in PE calculation, it is suggested to use the ancestral range A_i to indicate the true evolutionary history of species in that range. Assuming there is an internal node i (or ancestor species i) has the branch l_i which is one in the node set $N(s)$ that connect the root to the focusing tip species s. I quantify the weight of phylogenetic ancestral endemism as l_i / A_i for that internal branch. As such, when I calculate through all the weights for the nodes inside the node set $N(s)$ and sum them together to obtain phylogenetic ancestral endemism (PAE) index for the focused species s. The formula is written as,

$$PAE_s = \sum_{i \in N(s)} l_i / A_i \tag{2}$$

where A_i is the ancestral range size constructed for the internal node. Thus, my PAE is not directly associated to the contemporary distribution of species, but only the ancestral range information involved.

To incorporate evolutionary distinctiveness index (ED) efficiently [13, 14, 21], the above equation (2) could be further extended to the form as follows,

$$PAE_s = \sum_{i \in N(s)} l_i / (S_i \times A_i) \tag{3}$$

where S_i denotes the number of external species descended from the concerned internal node *i*.

The above equation could be utilized to evaluate the conservation priority of each of the external species in the phylogeny. When the distributional range of species is assumed to be identical, the above equation (3) is degenerated into the ED index [13, 14, 21]. For evaluating the conservation importance of different areas, one could simply sum the PAE values for the species found in each of the focused areas. In the subsequent analyses, I will utilize equation (3) to measure species' conservation values.

When the distributional range sizes for external species are unknown and only the distributional grids/quadrates of species are presented, one could still utilize the above equation (3). However, for this time the reconstruction of ancestral ranges must be carried out on each of distributional grids for the species. As such, the range size of a species is simply the summation of distributional grids where the species occurs. The range size for each of its ancestors is analogous by summing the occurrence probabilities of the focused ancestor across all the distributional grids.

A Practical Example

I here used an empirical dataset to demonstrate the usage of PAE and compare it with other phylogenetic diversity indices. The dataset comprises of the distribution of 11 endemic Psychotria plants in four island groups of Hawaii [22, 23], which are group A (Kauai and Niihau islands), group B (Oahu island), group C (Molokai, Lanai, Maui and Kahoolawe islands) and group D (Hawaii island). They were made from the distribution matrix of Psychotria species established in previous studies [22, 23], for which I used with a modification by merging the distribution of different populations of the same species together. The tree for the 11 endemic Psychotria plants is pruned from a full tree in which different populations of the same species have been sampled [22, 23]. The pruned tree and distribution of the species over the four island groups is showed in Fig. **2**.

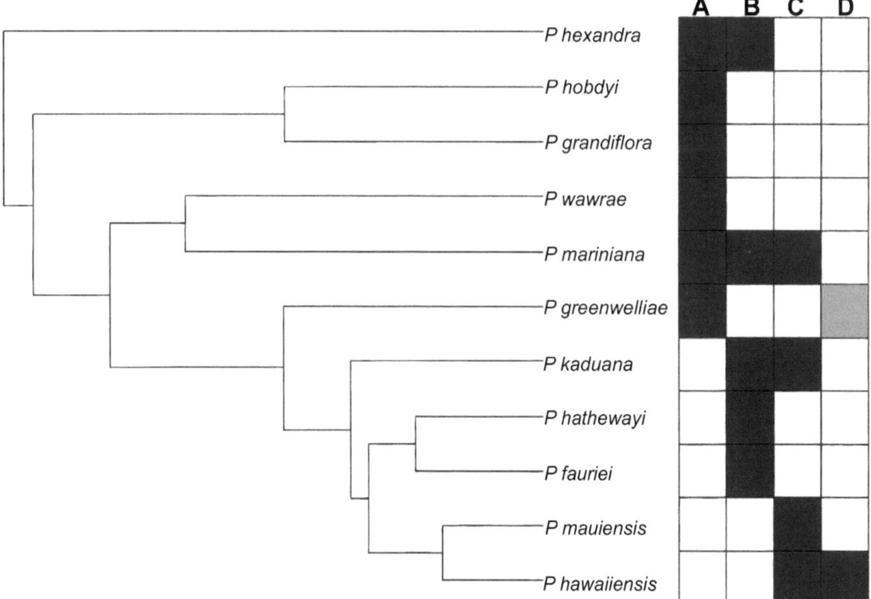

Figure 2: Phylogeny and distribution of 11 endemic Psychotria plants of Hawaii Islands. Different letters above the grids on the right-hand side indicated different island groups. They are group A (Kauai and Niihau islands), group B (Oahu island), group C (Molokai, Lanai, Maui and Kahoolawe islands) and group D (Hawaii island). Dark color in the squares indicated presence of the species in the island group, while white color indicated absence.

I compare PAE to other phylogenetic diversity indices, including PE, ED, taxonomic distinctiveness (TD) [15], I and W indices (node-based indices) [24].

Index PE is calculated using the equation (1), while index ED is calculated as follows,

$$ED_s = \sum_{i \in N(s)} l_i / S_i \tag{4}$$

where S_i is the number of external species descended from the internal node *i*.

Index TD is to replace the branch lengths by the integer 1 as,

$$TD_s = \sum_{i \in N(s)} 1 / S_i \tag{5}$$

Index I assigns a value of 1 to each terminal taxon that belongs to a pair of terminal sister taxa. The taxon that constitutes the sister group of this pair receives a value of 2 (equal to the sum of its sister group). Each successive taxon receives a value equal to that of the total sister group. Thus, index I refers to the number of phylogenetic groups to which a taxon belongs [24].

Index W measures the proportion that each taxon contributes to the total diversity of the group. Specifically, index W assigns an information value (i) to each terminal taxon. This value is calculated as the number of groups (nodes) to which each taxon belongs. A weight (Q) is calculated as follows:

$$Q_j = \sum i / i_j \tag{6}$$

Where j is equal to each specific taxon in the cladogram. The Q value for each taxon refers to the proportion of total diversity of the group that is contributed by this taxon. The PD measure (W) is calculated as:

$$W = Q_j / Q_{\min} \tag{7}$$

where $Q_{\min}$ refers to the lowest Q-value for the entire group.

For Hawaii Psychotria plant data set, because the distribution of each plant in the four islands is known, all indices were performed. For calculating PAE index, the ancestral distribution probability of each species in each of the four island groups is reconstructed using a maximum likelihood procedure with the "ace" function of "ape" package [25] under R environment [26]. The maximum likelihood method is to treat the ancestral ranges at internal nodes of the phylogenetic tree as parameters, and attempts to find the parameter values that maximize the probability of the data (the observed range sizes for external species) given the dated phylogeny and the hypothesis of the evolution model of the ancestral ranges over the evolutionary time.

In the present study, this simple reconstruction of ancestral ranges assumes equal evolutionary transition rates among different islands. The summation of the probabilities over the four island groups thus is used to represent the ancestral range size for PAE calculation. For all the phylogenetic diversity indices, a standardized procedure is applied (each value is divided by the maximal value for each index).

RESULTS

For Hawaiian Psychotria endemic plants, the standardized PAE, PE, ED, TD, I, and W for each species are presented in Fig. **3**. As shown, PAE and PE performed differently (Fig. **4A**), there is no a significant correlation between them. Further, PAE has a significant correlation with ED (Fig. **4B**), but the correlations between PAE and TD, I and W are not significant (Fig. **4C-E**). Not surprisingly, the significant correlation between ED and PAE is due to the fact that PAE calculated by using equation (3) explicitly incorporated the distinctiveness information of species as ED.

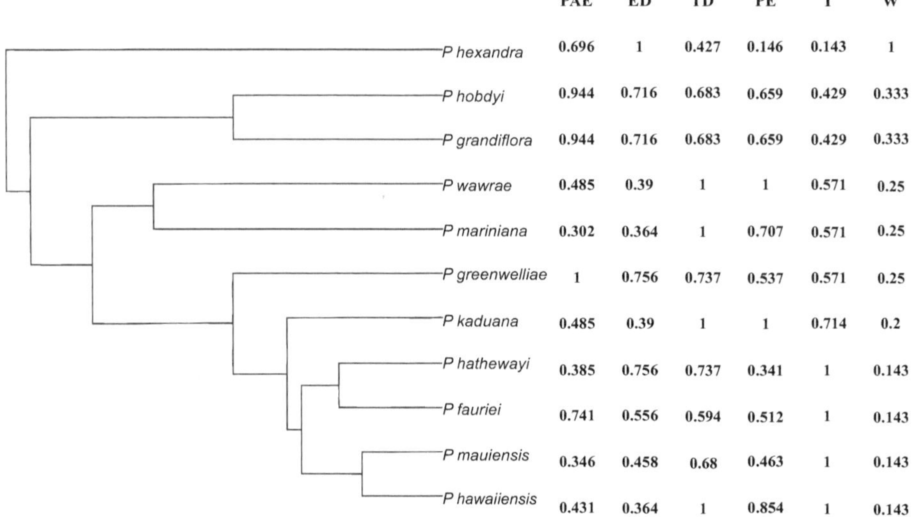

	PAE	ED	TD	PE	I	W
P hexandra	0.696	1	0.427	0.146	0.143	1
P hobdyi	0.944	0.716	0.683	0.659	0.429	0.333
P grandiflora	0.944	0.716	0.683	0.659	0.429	0.333
P wawrae	0.485	0.39	1	1	0.571	0.25
P mariniana	0.302	0.364	1	0.707	0.571	0.25
P greenwelliae	1	0.756	0.737	0.537	0.571	0.25
P kaduana	0.485	0.39	1	1	0.714	0.2
P hathewayi	0.385	0.756	0.737	0.341	1	0.143
P fauriei	0.741	0.556	0.594	0.512	1	0.143
P mauiensis	0.346	0.458	0.68	0.463	1	0.143
P hawaiiensis	0.431	0.364	1	0.854	1	0.143

Figure 3: Standardized PAE, PE, ED, TD, I, and W indices for the 11 endemic Psychotria plants of Hawaii Islands.

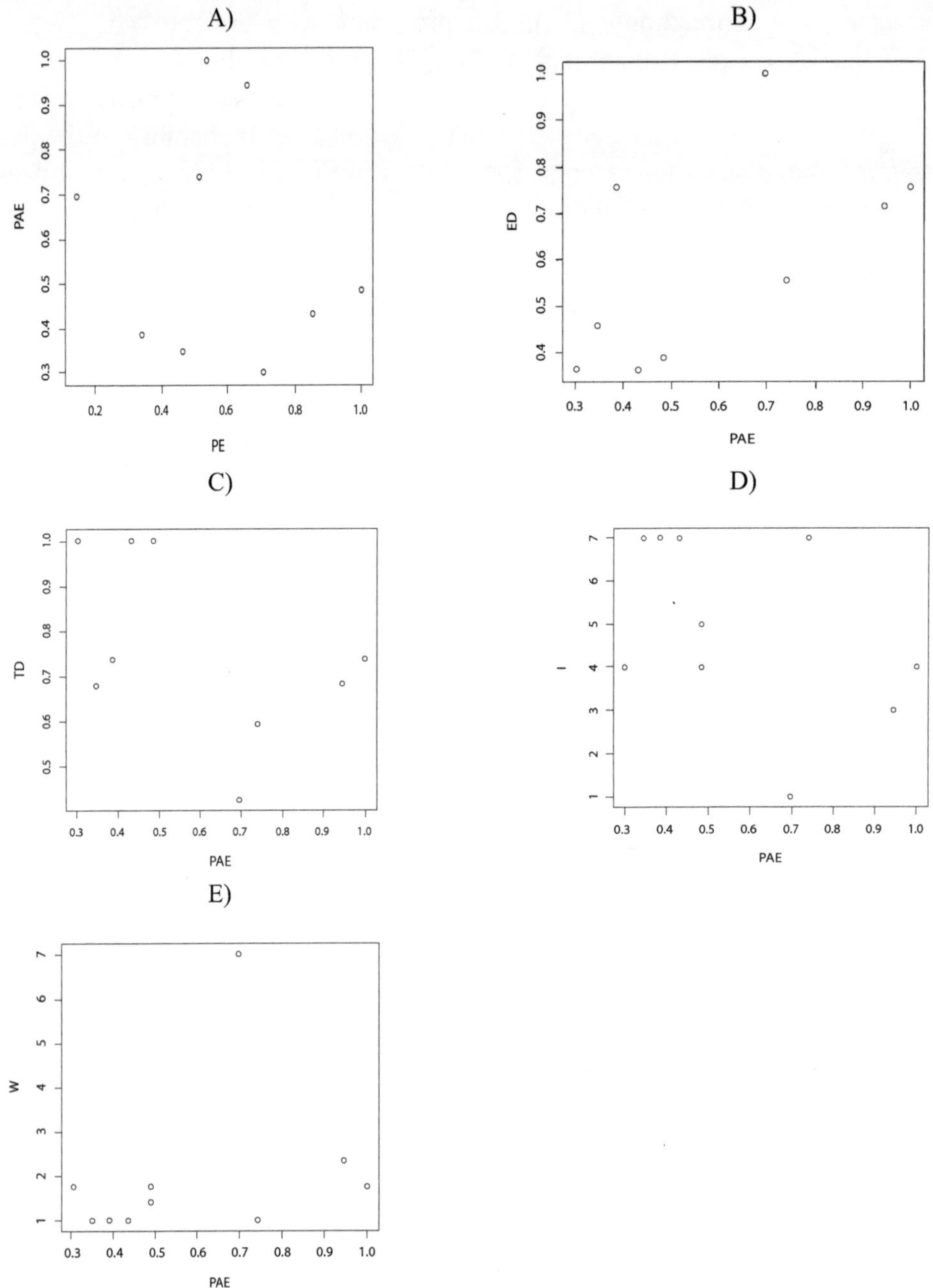

Figure 4: The relationships of PAE with other diversity indices (PE, ED, TD) for Psychotria endemic plants of Hawaii Islands. A) PAE *versus* PE (correlation = 0.16, P > 0.05); B) PAE *versus* ED (correlation = 0.6, P < 0.05); C) PAE *versus* TD (correlation = 0.51, P > 0.05); D) PAE *versus* I (correlation = -0.5, P > 0.05); E) PAE *versus* W (correlation = 0.28, P > 0.05).

DISCUSSION

PAE index effectively integrates the ED index [13, 14]. Therefore, this index actually incorporates both "taxonomic rarity" and "distributional endemism" of species with the consideration of range heritage [27, 28] over evolutionary history. In comparison to PE index, PAE didn't utilize the union of contemporary distribution of external species to measure the ranges of ancestors. As such, it didn't need to handle distributional overlapping status of species when performing the union of the ranges of different species in a specific clade. This is advantageous since PAE only considered the ancestral range size of species, but not the range sizes of individual species for their most common ancestors. As such, one could evaluate conservation priority of species in a simple way without knowing the exact distributional ranges of species, as long as the range size data are available.

PAE index is fundamentally different from other trait/range-weighted phylogenetic diversity indices as well, for example, abundance-weighted evolutionary distinctiveness index [29] or biogeographically weighted evolutionary distinctiveness index [30], because these previous studies take the union of or sum the ranges/abundance for descendants as did by PE index. As repeatedly discussed above, the ignorance of ancestral range information would actually lead to possibly biased results and could not reflect the true range heritage and evolution scenarios over the phylogenetic history. In contrast, the PAE index explicitly incorporates ancestral information of traits or ranges of species, thus accurately reflecting the "lineage endemism" over evolutionary time for either external survived species or internally extinct ancestors.

The importance of introducing ancestral ranges for setting conservation priority of external species using PAE index stems from the hypothesis of phylogenetic niche conservation [31, 32] which states that the descendants tend to retain similar traits as the ancestors over the evolutionary time. Thus, the estimation of ancestral range information for the external species when setting up conservation priority of species can allow ones to incorporate the important areas that have allowed many extinct ancestral lineages and external species to inhabit.

Consequently, those external species that have ancestors distributed in some distinct habitats/areas should be paid more attention in conservation given the fact that their genes that are inherited from ancestral lineages may contain some unique physiological and/or functional abilities to adapt to the distinct habitat conditions. At last, the important areas predicted by PAE index might be of great importance in conservation since they should be the ones for species diversification and the survival of newly speciated species in the future based on the prediction of phylogenetic niche conservatism hypothesis.

However, it is acknowledged that there are some potential limitations for the application of the new index PAE and the shortcoming of the present study: firstly, it requires accurate estimation of ancestral ranges of species over the phylogenetic time scale. Actually, the ancestral state reconstruction is still challenging because there are many uncertainties (and the trend is to utilize Bayesian principle to estimate ancestral states in the internal nodes in the tree). Thus, PAE index might become much uncertain if the ancestral range information is undetermined. Secondly, the present study utilizes only 11 endemic Psychotria plants of Hawaii Islands as an example for applying the index, which is thus largely not enough, especially when ones compare the differences between PAE and other indices using correlation analysis as Fig. **4**. As such, more empirical tests and applications of the PAE index and comparison with other phylogenetic diversity indices on other plant taxa should be carried out in the future.

In conclusion, the present study proposed a new phylogenetic diversity index PAE for potential applications in systematic conservation planning [33]. The new index explicitly takes into account the role of range heritage over evolutionary history by incorporating the information of ancestral distributional ranges of ancestors. The index combines evolutionary distinctiveness and range sizes, being potentially promising to indicate conservation importance of organisms and areas.

REFERENCES

[1] D. Faith, "Conservation evaluation and phylogenetic diversity", *Biol. Conserv.* 61, 1-10, 1992.

[2] D. Faith, C. Reid, J. Hunter, "Integrating phylogenetic diversity, complementarity, and endemism for conservation assessment", *Conserv. Biol.* 18, 255-261, 2004.

[3] D. Faith, "Threatened species and the potential loss of phylogenetic diversity: conservation scenarios based on estimated extinction probabilities and phylogenetic risk analysis", *Conserv. Biol.* 22, 1461-1470, 2008.

[4] D. Faith, "Quantifying biodiversity: a phylogenetic perspective", *Conserv. Biol.* 16, 248-252, 2002.

[5] J. Diniz-Filho, R. Loyola, P. Raia, A. Mooers, L. Bini, "Darwinian shortfalls in biodiversity conservation", *Trends Ecol. Evol.* 28, 689-695, 2013.

[6] O. Petchey, A. Hector, K. Gaston, "How do different measures of functional diversity perform?", *Ecology*. 85, 847-857, 2004.

[7] D. Faith, P. Walker, "Environmental diversity: on the best-possible use of surrogate data for assessing the relative biodiversity of sets of areas", *Biodivers. Conserv.* 5, 399-415, 1996.

[8] Y. Chen, "Combining the species-area-habitat relationship and environmental cluster analysis to set conservation priorities: a study in the Zhoushan Archipelago, China", *Conserv. Biol.* 23, 537-545, 2009.

[9] M. Araujo, C. Humphries, P. Densham, R. Lampinen, W. Hagemeijer, A. Mitchell-Jones, *et al.*, "Would environmental diversity be a good surrogate for species diversity?", *Ecography*. 24, 103-110, 2001.

[10] N. Cooper, R.P. Freckleton, W. Jetz, "Phylogenetic conservatism of environmental niches in mammals", *Proc. R. Soc. B Biol. Sci.* 278, 2384-91, 2011. doi:10.1098/rspb.2010.2207.

[11] S. Taylor, P. Franks, S. Hulme, E. Spriggs, P. Christin, E. Edwards, *et al.*, "Photosynthetic pathway and ecological adaptation explain stomatal trait diversity amongst grasses", *New Phytol.* 193, 387-396, 2012.

[12] Y. Chen, "Conservation priority of global Galliformes species based on phylogenetic diversity", *Integr. Zool.* 9, 340-348, 2014.

[13] D. Redding, A. Mooers, "Incorporating evolutionary measures into conservation prioritization", *Conserv. Biol.* 20, 1670-1678, 2006.

[14] D. Redding, K. Hartmann, A. Mimoto, D. Bokal, M. Devos, A. Mooers, "Evolutionarily distinctive species often capture more phylogenetic diversity than expected", *J. Theor. Biol.* 251, 606-615, 2008.

[15] M. Cadotte, T. Davies, "Rarest of the rare: advances in combining evolutionary distinctiveness and scarity to inform conservation at biogeographical scales", *Divers. Distrib.* 16, 376-385, 2010.

[16] Y. Chen, "A phylogenetic subclade analysis of range sizes of endemic woody see plant species of China: trait conservatism, diversification rates and evolutionary models", *J. Syst. Evol.* 51, 590-600, 2013.

[17] Y. Chen, "Modeling extinction risk of endemic birds of mainland China", *Int. J. Evol. Biol.* 2013 (2013) 639635.

[18] Y. Chen, "Conservation priority for endemic birds of mainland China based on a phylogenetic framework", *Chin. Birds.* 4, 248-253, 2013.

[19] D. Rosauer, S. Laffan, M. Crisp, S. Donnellan, L. Cook, "Phylogenetic endemism: a new approach for identifying geographical concentrations of evolutionary history", *Mol. Ecol.* 18, 4061-4072, 2009.

[20] R. Gudde, J. Joy, A. Mooers, "Imperiled phylogenetic endemism of Malagasy lemuriformes", *Divers. Distrib.* 19, 665-675, 2013.

[21] I. Martyn, T. Kuhn, A. Mooers, V. Moulton, A. Spillner, "Computing evolutionary distinctiveness indices in large scale analysis", *Algorithms Mol. Biol.* 7, 6, 2012.

[22] M. Nepokroeff, K. Sytsma, W. Wagner, E. Zimmer, "Reconstructing ancestral patterns of colonization and dispersal in the Hawaiian understory tree genus Psychotria (Rubiaceae): A comparison of parsimony and likelihood approaches", *2Systematic Biol.* 52, 820-838, 2003.

[23] R. Ree, S. Smith, "Maximum likelihood inference of geographic range evolution by dispersal, local extinction and cladogenesis", *Syst. Biol.* 57, 4-14, 2008.

[24] P. Posadas, D. Esquivel, J. Crisci, "Using phylogenetic diversity measures to set priorities in conservation: an example from southern South America", *Conserv. Biol.* 15, 1325-1334, 2001.

[25] E. Paradis, J. Claude, K. Strimmer, "APE: analyses of phylogenetics and evolution in R language", *Bioinformatics.* 20, 289-290, 2004.

[26] R Development Core Team, R: A Language and Environment for Statistical Computing, Vienna, Austria. ISBN 3-900051-07-0, URL http://www.R-project.org., (2013).

[27] T. Webb, K. Gaston, "On the heritability of geographic range sizes", *Am. Nat.* 161, 553-566, 2003.

[28] A. Machac, J. Zrzavy, D. Storch, "Range size heritability in Carnivora is driven by geographic constraints", *Am. Nat.* 177, 767-779, 2011.

[29] M. Cadotte, J. Davies, J. Regetz, S. Kembel, E. Cleland, T. Oakley, "Phylogenetic diversity metrics for ecological communities: integrating species richness, abundance and evolutionary history", *Ecol. Lett.* 13, 96-105, 2010.

[30] C. Tucker, M. Cadotte, T. Davies, T. Rebelo, "Incorporating geographical and evolutionary rarity into conservation prioritization", *Conserv. Biol.* 26, 593-601, 2012.

[31] J. Losos, "Phylogenetic niche conservatism, phylogenetic signal and the relationship between phylogenetic relatedness and ecological similarity among species", *Ecol. Lett.* 11, 995-1003, 2008.

[32] M. Crisp, L. Cook, "Phylogenetic niche conservatism: what are the underlying evolutionary and ecological causes?", *New Phytol.* 196, 681-694, 2012.

[33] C. Margules, R. Pressey, "Systematic conservation planning", *Nature.* 405, 243-253, 2000.

Chapter 16:Can Higher Taxonomic Hierarchy Units be Effective Surrogates of Plant Hotspots and Conservation Areas? A Test on Endemic Plants in a Tropical Biodiversity Hotspot

Biodiversity and Biogeographic Patterns in Asia-Pacific Region I: Statistical Methods and Case Studies, **2015, 133-141**

CHAPTER 16

Can Higher Taxonomic Hierarchy Units be Effective Surrogates of Plant Hotspots and Conservation Areas? A Test on Endemic Plants in a Tropical Biodiversity Hotspot

Abstract: In this study, we will test whether plant family and genus richness could represent regional-scale species richness patterns. We also test whether plant family and genus richness could identify hotspots and complementary priority areas for maximally conserving species. The distribution of 340 endemic plants in Western Ghats of India, a tropical biodiversity hotspot in South Asia was used as a case study. The results implied that the spatial richness patterns created from the two higher taxonomic hierarchy units (family and genus) could very effectively represent plant species richness hotspots. However, the complementary priority areas selected by family- and/or genus-based data were very different from those selected based on species-site matrix for endemic plants in Western Ghats region. In conclusion, family and genus seemed to be good surrogates to reflect species in mapping biodiversity hotspots but were in low efficiency in selecting complementary conservation priority areas to conserve as many species as possible. Our study should shed some interesting insights into rapid assessment of regional biodiversity and quick identification of conservation targets.

Keywords: Biodiversity hotspots, complementarity principle, conservation priorities, ecological communities, ecological indicators, original forest, phylogenetic affinity, plant richness, species association, species classification tree, surrogates, systematic conservation planning, taxonomic diversity, taxonomic hierarchy.

INTRODUCTION

Conservation biologists have used surrogate information as a shortcut to monitor or solve conservation issues for a long time [1]. The merging challenge posed accordingly is the identification of appropriate surrogates. However, the appropriateness of each kind of surrogate is controversial in scientific communities so far. It is hard to reach a consensus that which kind of surrogate would perform best in capturing biodiversity. In this study, we would focus on the higher-taxon approach in representing biodiversity status.

In many cases, we could not obtain enough plant species distributional information at the surveyed location. However, at least we might obtain sufficient information at higher taxonomic units. Therefore studying the ability of higher

Youhua Chen

taxonomic hierarchy units to represent regional biodiversity [2] will provide a possible alternative approach to understand regional biodiversity status and therefore set up conservation targets.

In this study, at the genus and family levels, we will test whether the complementarity analysis could capture biodiversity as at the species level and determine the factors influencing the robustness of HTCs as surrogates. The study should shed some insights into protecting regional biodiversity using surrogate information.

The meaning of using HTCs to select conservation targets is that it could decrease the identification time and satisfy the needs of experts [3]. As far as I know, the studies on evaluation of HTCs as surrogate of species diversity and conservation priorities were still limited up to date. Heino and Soininen [4] discussed the utility of families and genera to serve as surrogate in capturing biodiversity information in the stream organisms, which suggested that genus-level surrogate would perform better than family-level surrogate.

Further, it has not been discussed completely so far under which circumstances employing HTCs as surrogates are suitable. In this study, we will employ the available information from genera and family to rigorously address this unsolved issue. Our study has two objectives: (1) the ability of HTCs to richness mapping and to identify complementarily priority areas for conservation; (2) the determinants of applying HTCs as surrogates. For brevity, we only set HTCs to genus and family levels.

MATERIALS AND METHODS

Data Sets

The distributional records of 341 endemic trees in Western Ghats of India are collected from the India Biodiversity Portal (http://indiabiodiversity.org/). Species checklist is available upon request. 0.25 latitude $\times$ 0.25 longitude spatial grain is applied [5] to obtain sampling grid cells for these species, resulting into 172 grid cells (Fig. **2**).

Methods

Richness Mapping and Spatial Correlation

To map regional diversity, the number of plant family, genus and species at each county was counted. The software ArcView v3.3 was used to map the richness.

To assess the effectiveness of family and genus in representing species diversity, the comparison of similarity for species richness, family richness and genus richness vectors should be carried out. In the present study, the spatial correlation test was used to compare vector similarity and thus determine whether family- and genus-level richness patterns could be used as surrogates of species richness spatial patterns. The purpose of utilizing spatial correlation test is to reduce the influence of spatial autocorrelation on the true association between species, family and genus richness patterns.

Computation of spatial correlation test was done as follows. For two spatial sequences $\{x\}$ and $\{y\}$, the length of which is N (*i.e.*, there are N sampling geographic locations). A way to adjust the sample size is as follows [6, 7],

$$M = 1 + \frac{N^2 \sigma_x^2 \sigma_y^2}{trace\{W_x W_y\}} \tag{1}$$

where M is the adjusted sample size. σ_x^2 and σ_y^2 are the variance of the sequences $\{x\}$ and $\{y\}$ respectively. W_x and W_y are the variance-covariance matrix for both spatial sequences $\{x\}$ and $\{y\}$, which can be computed using variogram method.

After obtaining the adjusted sample size, the significance of the Pearson's simple correlation r_{obs} is calculated using the following formula [6, 7],

$$p = 1 - \int_{-r_{obs}}^{r_{obs}} f(r)dr \tag{2}$$

$$\text{Here, } f(r) = \frac{(1-r^2)^{(M-4)/2}}{Beta(0.5, M/2-1)} \tag{3}$$

Conservation Priority Selection

Complementarity analysis was employed to select priority areas for conserving endemic plants. We utilize both rarity and richness criteria to select complementary priority areas. Complementarity analysis was conducted in iterative ways. In detail, for each iteration, the site with highest rarity or richness would be selected and the corresponding species that were found in the site and the site were removed from the species-site matrix. The same criterion was applied to next step of iteration until there were no species left in the system. For rarity index computed for a site, it is computed as,

$$Rarity = \sum_{i=1}^{n} I(i) \times 1 / s_i \tag{4}$$

Where s_i is the occurrence times of species *i* over all the sites, *I(i)* is an indicator function and becomes 1 if species i is presented in the site. *n* denotes the total number of species across all the sites.

RESULTS

Richness Mapping

The corresponding species-, genus- and family-level spatial richness patterns were shown in Figs. **1-3** respectively. As seen, there was a high similarity between these richness maps by visual check: all spatial patterns presented high richness in southern part of the region, while low richness in northern part of the region. Simple correlation maps were presented in Fig. **4**. As seen, the pairs between species richness, genus richness and family richness over the sampling grid cells at 0.5 latitude × 0.5 longitude were highly correlated.

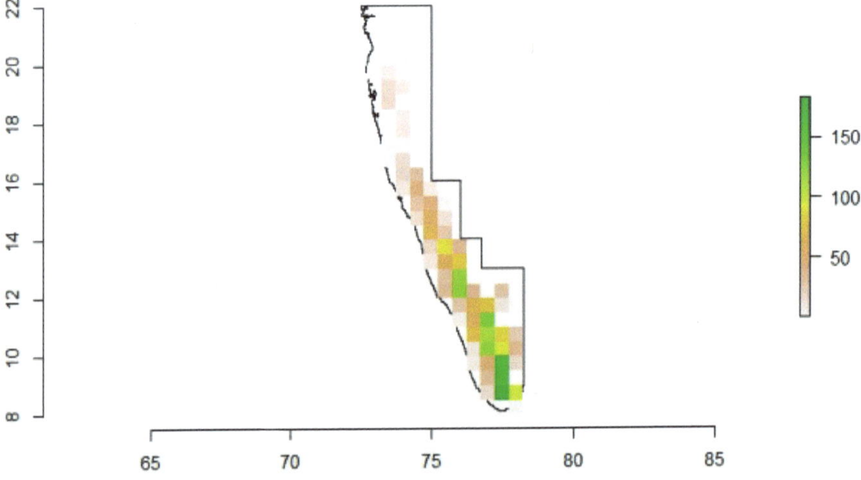

Figure 1: Richness mapping of endemic tree species over the sampling quadrats across Western Ghats of India with the spatial resolution of 0.5 latitude × 0.5 longitude. Colors from green to white indicate the species number from high to low for each grid cell. The x-axis of the map indicates the longitude (E), while the y-axis indicates the latitude (N).

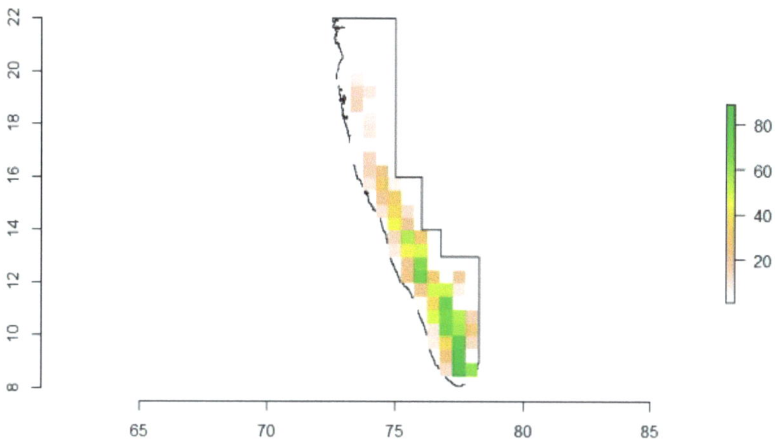

Figure 2: Richness mapping of endemic tree genera over the sampling quadrats across Western Ghats of India with the spatial resolution of 0.5 latitude × 0.5 longitude. Colors from green to white indicate the genera number from high to low for each grid cell. The x-axis of the map indicates the longitude (E), while the y-axis indicates the latitude (N).

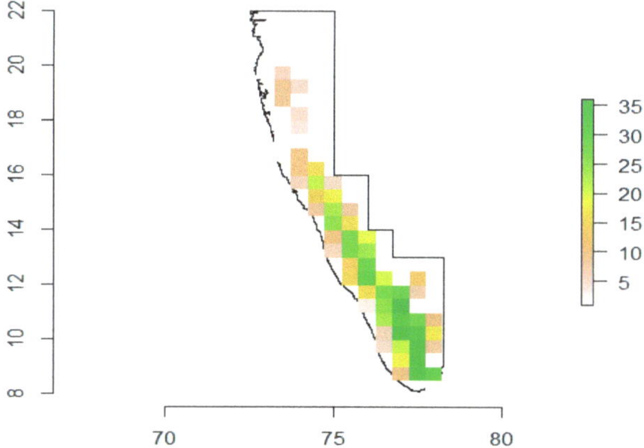

Figure 3: Richness mapping of endemic tree families over the sampling quadrats across Western Ghats of India with the spatial resolution of 0.5 latitude × 0.5 longitude. Colors from green to white indicate the family number from high to low for each grid cell. The x-axis of the map indicates the longitude (E), while the y-axis indicates the latitude (N).

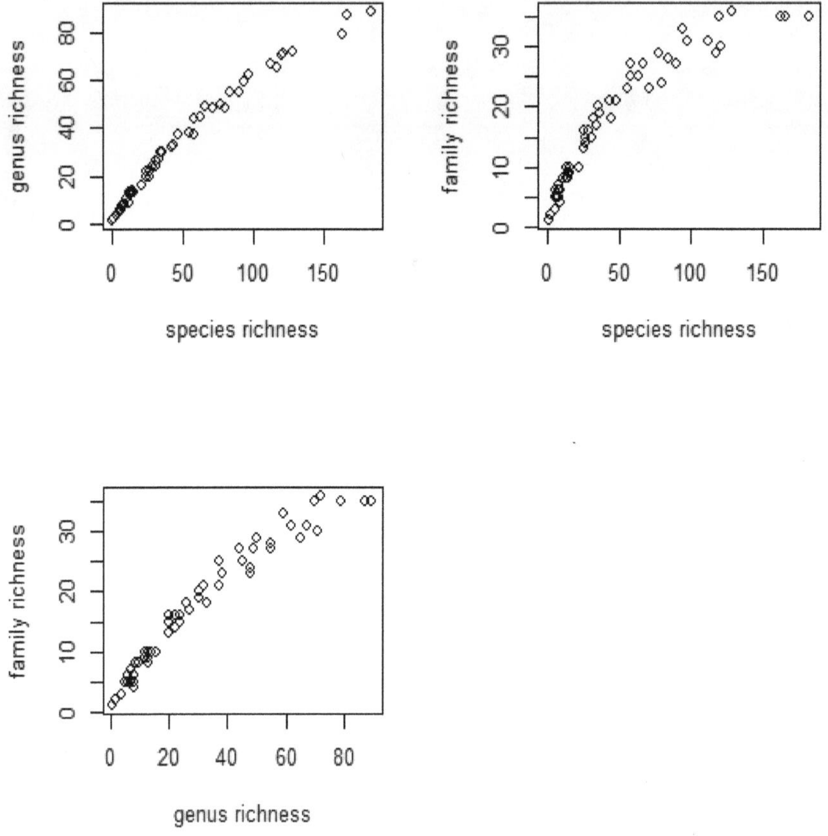

Figure 4: Correlations between species-, genus-, and family-richness across sampling grid cells.

Spatial correlation test further showed that, the correlation between species and genus richness was statistically significant even after removing spatial autocorrelation ($r=0.984$, $F=496.07$, $df=15.91$, $p<0.001$); the correlation between family and genus richness was statistically significant even after removing spatial autocorrelation ($r=0.973$, $F=239.95$, $df=13.64$, $p<0.001$); the correlation between species and family richness was statistically significant even after removing spatial autocorrelation ($r=0.926$, $F=91.25$, $df=15.06$, $p<0.001$).

Complementary Priority Areas

We numbered the sampling grid cells from top left ones to bottom right ones (Fig. **5**), for the purpose to conduct complementarity analysis to select priority areas for conservation.

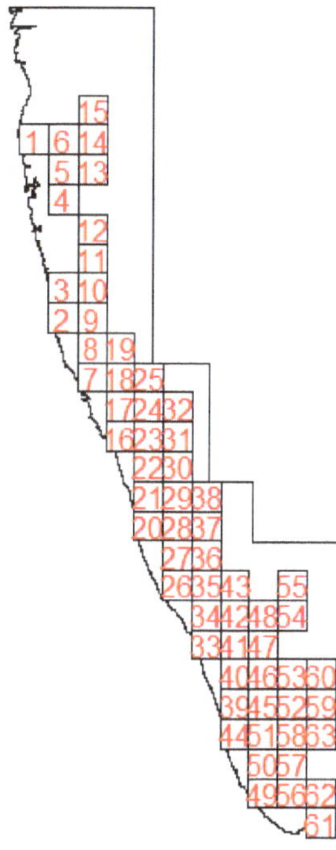

Figure 5: Numbered grid cells used for complementarity analysis to select priority areas.

At species level, for covering all the 340 endemic plant species based on richness criterion, the required complementary areas set contained 17 grid cells, which were 56, 47, 58, 29, 57, 36, 48, 42, 52, 35, 18, 53, 22, 23, 28, 38, 39, 41, 43, 45, 50, 51, 59, and 62. In comparison, based on rarity criterion using equation (4), the relevant complementary priority areas set contained the following sites in order: 56, 57, 47, 58, 36, 42, 48, 18, 52, 35, 29, 23, 28, 38, 39, 41, 43, 45, 50, 51, 53, 62, 22, and 59.

At genus level, for covering all the 118 genera that the species have, the required complementary areas set contained the grid cells in decreasing orders as follows: 56, 47, 57, 22, 62, 38, 48, and 53. In comparison, when rarity index was applied (equation (4)), the corresponding priority area set had the sites 56, 57, 48, 38, 62, 53, 22 and 35.

At family level, for covering the 43 families that the endemic plants have, the required complementary set contained 3 cells, which were 47, 45 and 50 respectively. In contrast, when rarity criterion was used, the priority area set for conserving families was composed of sites 57 and 48.

DISCUSSION

Plant family and genus could successfully represent species diversity, they could be served as the surrogate of regional biodiversity and provide a fairly effective approach to assess regional plant diversity. Based on our empirical observation, I suggest using family and genus information to map plant biodiversity and identify potential priority areas.

It is wise to employ surrogate to select conservation targets in the event of detailed species distribution information is lacking. The present study represented a step to realize the above principle. As a matter of fact, there are many other surrogates being proposed to solve the problem caused by uncompleted species distribution. Basically they could be classified into two categories: physical (or environmental) and biological ones [8]. For example, Pinto *et al.*, [9] employed bird richness to represent other taxonomic groups' richness and they have assessed the effectiveness of such a surrogate. However, there are some controversial voices to refute the surrogate approach. For example, Tognelli *et al.*, [10] argued that the spatial congruence of priority areas for different surrogates is extremely low.

Our study partially supported and refuted the previous studies: when richness hotspots were analyzed, families or genera hotspots did present similar spatial patterns as those species richness hotspots (Figs. **1-4**). Moreover, the spatial correlation test strongly supported that family or genus hotspots could be surrogates for species richness hotspots.

However, our study also supported the arguments made by Tognelli *et al.*, [10]: the complementary conservation priority areas selected for families, genera and species were actually very different regardless the criterion used (richness or rarity). Thus, there was no spatial congruence of priority areas at complementary respectively for the endemic plants of Western Ghats of India.

Finally, it should be noted that the study might be not well suited to apply to other taxonomic groups or spatial scales, for example, invertebrate and vertebrate taxonomic groups and at very local scales. However, because of lacking of relevant studies, further studies are wanted to better quantify the effectiveness of HTCs in conservation priority planning.

REFERENCES

[1] T. Caro, G. O'Doherty, "On the use of surrogate species in conservation biology", *Conserv. Biol.* 13, 805-814, 1999.

[2] M. Perez-Losada, A. Crandall, "Can taxonomic richness be used as a surrogate for phylogenetic distinctness indices for ranking areas for conservation?", *Anim. Biodivers. Conserv.* 26, 77-84, 2003.

[3] Y. Mandelik, T. Dayan, V. Chikatunov, V. Kravchenko, "Reliability of a higher-taxon approach to richness, rarity, and composition assessments at the local scale", *Conserv. Biol.* 21, 1506-1515, 2007.

[4] J. Heino, J. Soininen, "Are higher taxa adequte surrogates for species-level assemblage patterns and species richness in stream organisms?", *Biol. Conserv.* 137, 78-89, 2007.

[5] D. Currie, "Energy and large-scale patterns of animal- and plant-species richness", *Am. Nat.* 137, 27-49, 1991.

[6] P. Clifford, S. Richardson, D. Hemon, "Assessing the significance of the correlation between two spatial processes", *Biometrics.* 45, 123-134, 1989.

[7] P. Dutilleul, "Modifying the t test for assessing the correlation between two spatial processes", *Biometrics.* 49, 305-314, 1993.

[8] Y. Carmel, L. Stoller-Cavari, "Comparing environmental and biological surrogates for biodiversity at a local scale", *Isreal J. Ecol. Evol.* 52, 11-27, 2007.

[9] M. Pinto, J. Diniz-Filho, L. Bini, D. Blamires, T. Rangel, "Biodiversity surrogate groups and conservation priority areas: birds of the Brazilian Cerrado", *Divers. Distrib.* 14, 78-86, 2008.

[10] M. Tognelli, C. Silva-Garcia, F. Labra, P. Marquet, "Priority areas for the conservation of costal marine vertebrates in Chile", *Biol. Conserv.* 126, 420-428, 2005.

Chapter 16: Spatial Risk Assessment of Alien Plants in China Using Biodiversity Resistance Theory

Biodiversity and Biogeographic Patterns in Asia-Pacific Region I: Statistical Methods and Case Studies, **2015**, 143-148 143

CHAPTER 17

Spatial Risk Assessment of Alien Plants in China Using Biodiversity Resistance Theory

Abstract: In the present chapter, the potential occurrence risk of invasive plants across different provinces of China is studied using disease risk mapping techniques (empirical Bayes smoothing and Poisson-Gamma model). The biodiversity resistance theory which predicts that high-biodiversity areas will have reduced risk of species invasion serves as the base for performing spatial risk assessment of plant invasion across provinces. The results show that, both risk mapping methods identified that north-eastern part of China have the highest relative risk of plant invasion. In contrast, south-western and south-eastern parts of China, which have high woody plant richness, are predicted to possess low relative risks of plant invasion. For the future, it is interesting to compare the spatial risk patterns across different functional groups of these alien plants. The results of this chapter has been published in an open-access journal [1].

Keywords: Alien plants *versus* native plants, dynamics of invasiveness, ecological assemblages, habitat heterogeneity, invasion biology, invasive plants, plant distribution, risk assessment, spatial ecology, spatial regression analysis, spatial statistics, species distribution and diversity patterns, water availability.

INTRODUCTION

Alien plants may cause severe ecological problems onto local ecosystems because they can better compete for the resources than native species. Biodiversity hotspots are believed to be more resistant to species invasion: more diverse communities would have less species invasions [2, 3]. This biodiversity resistance theory can be an option for performing risk assessment of species invasions.

Modeling plant invasion risk in China has been done in some previous studies recently [4-7]. However, all these studies only rely on the richness information of alien plants, the dependence of alien plant diversity and native plant diversity has never been considered yet. As such, one may utilize the biodiversity resistance theory to quantify relative risk of alien plants by explicitly incorporating the information of native plant diversity.

In the present study, we perform the relative risk assessment of alien plants in China by using spatial disease mapping methods (including empirical Bayes smoothing and Poisson-Gamma model) so as to incorporate richness information of native woody plant species.

Youhua Chen

MATERIALS AND METHODS

Invasive Plant Diversity Data

The province-level 127 alien plant richness data are collected from previous studies [6, 8-10]. The resultant data include provincial distributional information and physiological trait patterns of each invasive plant found in China.

Woody Plant Diversity Data

The population size at risk when performing risk mapping is required as an entry. In the present study, we regard the population size at risk as the number of native woody plant species. As such, similar to the disease transmission model, areas where there are higher population sizes would have less relative risk of disease outbreak. In our modeling of relative risk estimation of plant invasion, we hold the recognition that areas with high biological diversity should have less plant invasion risk [2, 3].

Risk Mapping Using Empirical Bayes Smoothing and Poisson-Gamma Model

I use both empirical Bayes smoothing and Poisson-Gamma model for predicting the outbreak risk of invasive plants over the provinces of China. For testing whether the relative risk of plant invasion is significantly higher than 1, we utilize the Poisson exact test.

RESULTS AND DISCUSSION

As showed in Fig. **1**, the woody plant diversity is the highest in the southern part of China, which was identified as one of the global biodiversity hotspots of the world [11, 12]. Yunnan has the highest woody plant richness, followed by the Guangxi Province.

As showed in Fig. **2**, the alien plant diversity is highest in southern and eastern part of China. Interestingly, the biodiversity hotspot, Yunnan Province, also has very high alien plant diversity following Jiangsu Province [9].

The relative risk of plant invasion based on empirical Bayes smoothing and Poisson-Gamma models are highly similar (Figs. **3-4**). As seen, northern and western parts of China have high risk of plant invasion (dark grey colors). In contrast, southern part of China is less sensitive to plant invasion (light grey colors). Using Poisson exact test, it is identified that 13 provinces from northern part of China have significant relative risk values higher 1 (those provinces with red boundaries in Figs. **3-4**).

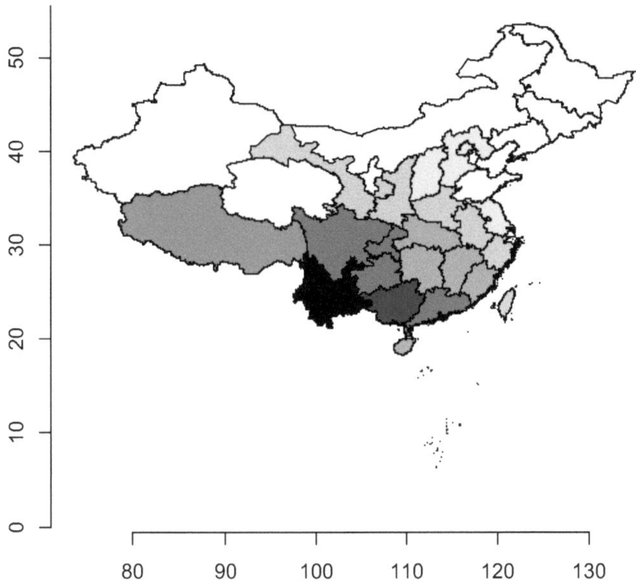

Figure 1: Observed woody plant diversity at provincial level of China. Colors from light to dark grey indicate diversity from low to high.

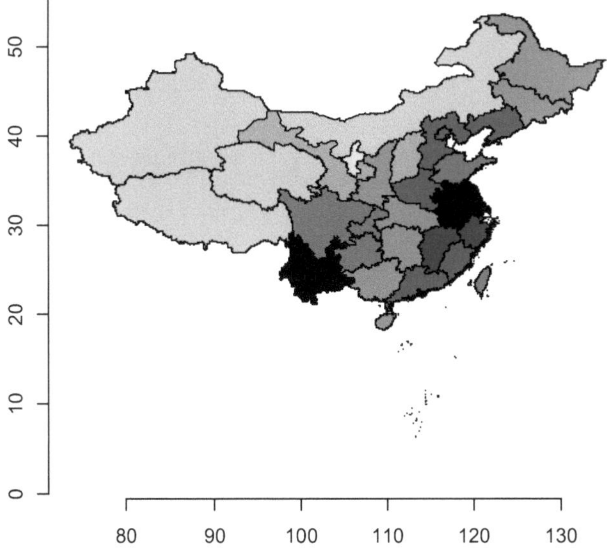

Figure 2: Observed alien plant diversity at provincial level of China. Colors from light to dark grey indicate diversity from low to high.

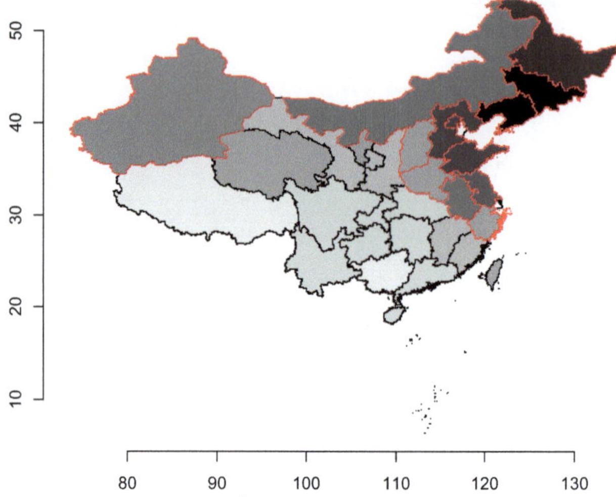

Figure 3: Relative risk of plant invasiveness at provincial level of China using empirical Bayes smoothing method. Provinces which have red boundaries indicate that their relative risks of plant invasion are significantly higher than 1. Colors from light to dark grey indicate invasive risks from low to high.

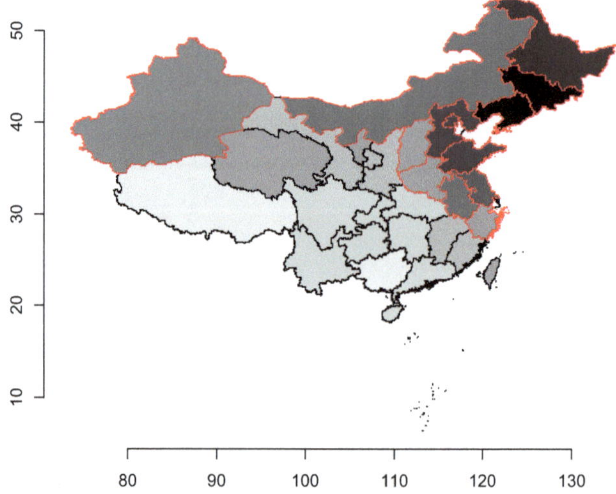

Figure 4: Relative risk of plant invasiveness at provincial level of China using Poisson-Gamma model. Provinces which have red boundaries indicate that their relative risks of plant invasion are significantly higher than 1. Colors from light to dark grey indicate invasive risks from low to high.

Different from previous studies, which showed that areas with high relative outbreak risk of alien plants are typically those with high richness of alien plants, the present study found that those areas with low native plant diversity would have higher risk of alien plant occurrence. The remarkable difference on the risk assessment is due to the inclusion of biodiversity information from native woody plant species. As mentioned above, all the previous studies [4-7] did not explicitly consider the interaction between native and alien plant diversity. The biodiversity resistance hypothesis [2, 3] provides a way to effectively incorporate diversity information from native plants for quantifying invasive risk of alien plants over provinces.

The present study evaluated the spatial risk assessment of alien plants as a whole. It would be interesting to study different functional groups of alien plants. Different functional groups may present different distributional patterns at provincial perspectives and the driving ecological factors might vary across different functional groups because of their differentiated traits and different functional responses mechanisms to the environmental change. Functional groups have been widely used in plant ecology studies [9, 13-15], thus, it would be worthy to evaluate the differentiation of spatial invasive risks of different functional groups of alien plants found in China.

REFERENCES

[1] Y. Chen, "Spatial risk assessment of alien plants in China using biodiversity resistence theory", *Comput. Ecol. Softw.* 4, 82-88, 2014.

[2] R. Law, R. Morton, "Permanence and the assembly of ecological communities", *Ecology.* 77, 762-775, 1996.

[3] J. Stachowicz, H. Fried, R. Osman, R. Whitlatch, "Biodiversity, invasion resistance, and marine ecosystem function: reconciling pattern and process", *Ecology.* 83, 2575-2590, 2002.

[4] F. Bai, R. Chisholm, W. Sang, M. Dong, "Spatial risk assessment of alien invasive plants in China", *Environ. Sci. Technol.* 47, 7624-7632, 2013.

[5] J. Liu, S. Liang, F. Liu, R. Wang, M. Dong, "Invasive alien plant species in China: regional distribution patterns", *Divers. Distrib.* 11, 341-347, 2005.

[6] X. Wu, J. Luo, J. Chen, B. Li, "Spatial patterns of invasive alien plants in China and its relationship with environmental and anthropological factors", *J. Plant Ecol.* 30, 576-584, 2006.

[7] J. Feng, Y. Zhu, "Alien invasive plants in China: risk assessment and spatial patterns", *Biodivers. Conserv.* 19, 3489-3497, 2010.

[8] Y. Chen, "Distributional patterns of alien plants in China: the relative importance of phylogenetic history and functional attributes", *ISRN Ecol.* 2013, 527052, 2013.

[9] Z. Wang, Y. Chen, Y. Chen, "Functional grouping and establishment of distribution patterns of invasive plants in China using self-organizing maps and indicator species analysis", *Arch. Biol. Sci.* 61, 71-78, 2009.

[10] X. Yan, H. Shou, J. Ma, "The problem and status of the alien invasive plants in China", *Plant Divers. Resour.* 34, 287-313, 2012.

[11] N. Myers, R. Mittermeier, C. Mittermeier, G. da Fonseca, J. Kent, "Biodiversity hotspots for conservation priorities", *Nature.* 403, 853-858, 2000.

[12] Y. Chen, J. Bi, "Biogeography and hotspots of amphibian species of China: Implications to reserve selection and conservation", *Curr. Sci.* 92, 480-489, 2007.

[13] W. Liu, R. Zang, Y. Ding, W. Zhang, "Species-area relationships of various plant functional groups in tropical monsoon rain forest (Hainan Island, China)", *Pol. J. Ecol.* (2013) In press.

[14] K. Steinmann, H. Linder, N. Zimmermann, "Modellling plant species richness using functional groups", *Ecol. Model.* 220, 962-967, 2009.

[15] J. Zhang, F. Zhang, "Diversity and composition of plant functional groups in mountain forests of the Lishan Nature Reserve, North China", *Bot. Stud.* 48, 339-348, 2007.

Index

E

F

G

H

I

Q

R

S

T

U

V

W

Z

www.ingramcontent.com/pod-product-compliance
Lightning Source LLC
Chambersburg PA
CBHW041709210326
41598CB00007B/592

* 9 7 8 1 6 8 1 0 8 0 1 6 1 *